Synthetic Sorbent Materials Based on Metal Sulphides and Oxides

Synthetic Sorbent Materials Based on Metal Sulphides and Oxides

D.S. Sofronov, K.N. Belikov, M. Rucki,
S.N. Lavrynenko, Z. Siemiątkowski,
E. Yu. Bryleva, and O.M. Odnovolova

Taylor & Francis
Taylor & Francis Group

First edition published 2021
by Taylor & Francis
6000 Broken Sound Parkway NW, Suite 300, Boca Raton, FL 33487-2742

and by Taylor & Francis
2 Park Square, Milton Park, Abingdon, Oxon, OX14 4RN

© 2021 Taylor & Francis Group, LLC

Reasonable efforts have been made to publish reliable data and information, but the author and publisher cannot assume responsibility for the validity of all materials or the consequences of their use. The authors and publishers have attempted to trace the copyright holders of all material reproduced in this publication and apologize to copyright holders if permission to publish in this form has not been obtained. If any copyright material has not been acknowledged, please write and let us know so we may rectify in any future reprint.

The Open Access version of this book, available at www.taylorfrancis.com, has been made available under a Creative Commons Attribution-Non Commercial-No Derivatives 4.0 license.

With the exception of the Introduction, Chapter 1.1, Final Remarks, and References no part of this book may be reprinted or reproduced or utilised in any form or by any electronic, mechanical, or other means, now known or hereafter invented, including photocopying and recording, or in any information storage or retrieval system, without permission in writing from the publishers.

The Introduction, Chapter 1.1, Final Remarks, and References of this book are available for free in PDF format as Open Access from the individual product page at www.routledge.com. It has been made available under a Creative Commons Attribution-Non Commercial-No Derivatives 4.0 license.

Trademark notice: Product or corporate names may be trademarks or registered trademarks, and are used only for identification and explanation without intent to infringe.

ISBN: 978-0-367-56675-3 (hbk)
ISBN: 978-0-367-60875-0 (pbk)
ISBN: 978-1-003-10233-5 (ebk)

Typeset in Palatino
by codeMantra

Contents

Authors vii
Acknowledgments ix
Abbreviations and Nomenclature xi

Introduction 1

1. Synthesis of Zinc, Copper, Cadmium, and Iron Sulfides and Their Sorption Properties 5
   1.1 Methodology of Synthesis 5
       1.1.1 Overview and Classification of Synthesis Methods 5
           1.1.1.1 High-Temperature Methods 6
           1.1.1.2 Solvothermal Decomposition Method 8
       1.1.2 Methodology and Equipment 11
       1.1.3 Procedures and Reactions 13
   Acknowledgment 14
   1.2 Precipitation with Sodium Sulfide 15
   1.3 Particles Precipitation Out of Thiourea Solutions 16
       1.3.1 Effect of Chemical Composition of the Reagent Environment on the Particle Formation of Zinc, Copper, and Cadmium Sulfides 16
           1.3.1.1 pH effect 16
           1.3.1.2 Concentration Effect 23
           1.3.1.3 Effect of the Anion Nature 26
       1.3.2 Impact of the Organic Additives 29
   1.4 Effect of Microwave Activation on the Zinc, Cadmium, and Copper Sulfide Particles Formation 37
       1.4.1 Effect of pH and Thiourea Concentration Combined with Microwave Activation 37
       1.4.2 Temperature Effect 42
   1.5 Effect of Metal Ion Doping on the Zinc Sulfide Particles Formation 44
   1.6 Iron Sulfide FeS Formation 54
   1.7 Sorption Properties of the Metal Sulfides 55
       1.7.1 pH Impact on the Metal Sulfides Extraction Efficiency and Particles Stability 56
       1.7.2 Impact of Specific Surface on the Sorption Capacity of Particles 58
       1.7.3 Sorption Properties of Zinc Sulfide Particles Doped with Copper, Manganese, and Cerium 60
   1.8 Conclusions on the Sorption Efficiency of Metal Sulfides 66

2. Formation and Sorption Properties of Iron Oxides and Manganese Oxyhydroxide 67
- 2.1 Synthesis of Iron Oxide 69
  - 2.1.1 Synthesis of the Hematite Particles α-Fe_2O_3 69
  - 2.1.2 Formation of Maghemite γ-Fe_2O_3 Particles 74
  - 2.1.3 Peculiarities of the Fe_3O_4 Particle Formation from the Aqueous Solutions 75
- 2.2 Synthesis of Manganese Dioxide and Oxyhydroxide 80
  - 2.2.1 Synthesis of Manganese Oxyhydroxide in Alkaline Solutions 80
  - 2.2.2 Synthesis of Manganese Dioxide 87
- 2.3 Sorption Characteristics of Iron Oxides and Manganese Dioxide and Oxyhydroxide 88
  - 2.3.1 Iron Oxides 88
  - 2.3.2 Manganese Oxyhydroxide and Dioxide 91
- 2.4 Conclusions 97

Final Remarks 99

References 101

Index 113

Authors

D.S. Sofronov received his PhD in inorganic chemistry in 2008. He was a younger researcher of State Scientific Institution, Institute for Single Crystals, National Academy of Sciences of Ukraine, Kharkov (Ukraine), from 2002–2014. Since 2014, he has been a researcher of State Scientific Institution, Institute for Single Crystals, National Academy of Sciences of Ukraine, Kharkov (Ukraine).
ORCID 0000-0003-4835-7001

K.N. Belikov received his PhD in analytical chemistry in 2002. He was deputy general director of the State Scientific Institution, Institute for Single Crystals, National Academy of Sciences of Ukraine, Kharkov (Ukraine), in 2015.
ORCID 0000-0002-1682-6064

M. Rucki received Polish Government Scholarship for PhD Studies, Poznan University of Technology (Poland), from 1994–1996. He obtained a PhD in metrology and measurement systems in 1997. He was an assistant professor at Poznan University of Technology, Institute of the Mechanical Engineering, Division of the Metrology and Measurement Systems, from 1997–2010. At present, he is working as a professor at Kazimierz Pulaski University of Technology and Humanities in Radom (Poland).
ORCID 0000-0001-7666-7686

S.N. Lavrynenko was a researcher at Kharkov State Polytechnic University from 1983–1994. He worked as an associate professor at National Technical University "KhPI" from 1994–2012. He graduated in 2001 from NATO Advanced Study Institute, "Responsive Systems for Active Vibration Control" at Universite Libre de Bruxelles, Faculte des Sciences, Brussels, Belgium. He was full professor at National Technical University "KhPI," Information Technology & Systems, KGM Department from 2012–2018. He was a Senior Visiting Fellow of the DAAD, Otto-von-Guericke University Magdeburg, in 1997. He was an Honorary Research Fellow of the Royal Society at University College London, Optical Science Lab, from 1999–2000. He was author and coauthor of more than 200 papers, reports, and books. He passed away on March 7, 2018, before this book was finished.
ORCID 0000-0003-2229-4858

Z. Siemiatkowski received his PhD in mechanical engineering and technology in 2000. In 2000 he became assistant at Kazimierz Pulaski University of Technology and Humanities (UTH) in Radom (Poland), Mechanical Faculty.

He was vice dean of Mechanical Faculty from 2012–2019, awarded with several individual awards for organizational and scientific achievements. Since 2019, he has been the head of the Division of Mechanical Engineering, UTH, Radom.

ORCID 0000-0002-6830-4479

E.Yu. Bryleva received his PhD in physical chemistry in 2008. He was a researcher at the State Scientific Institution, Institute for Single Crystals, National Academy of Sciences of Ukraine, from 2008–2013. Since 2013, he has been a senior researcher at the State Scientific Institution, Institute for Single Crystals, National Academy of Sciences of Ukraine.

ORCID 0000-0002-8903-4922

O.M. Odnovolova graduated from the State Scientific Institution Institute for Single Crystals, National Academy of Sciences of Ukraine in 2016. Since 2016, he has been a senior chemist in the analytical control sector of the chemical laboratory of the quality department, Kievmedpreparat, Kharkov, Ukraine.

ORCID 0000-0003-3497-1481

Acknowledgments

The authors express their gratitude for cooperation to everybody who contributed to the research, especially to chemistry doctors Baumer V.N. and Puzan A.N. for X-ray structural analysis, to Bunina Z.Yu., Gudzenko L.V., and chemistry doctor Shcherbakov I.B.-Kh. for chemical analysis, to Mateichenko P.V. for microstructural analysis, to chemistry doctor Beda A.A. for specific surface determination, and to physics and mathematics doctor Katrunov K.A. for optical researches.

The Open Access was sponsored through funding by Precision Machine Parts Poland Sp. z o.o.

Abbreviations and Nomenclature

Abbreviations

BET	Brunauer–Emmett–Teller model
CVD	chemical vapor deposition
ICP-AES	inductively coupled plasma atomic emission spectroscopy
IR	infrared
MAC	maximal acceptable concentration
MOCVD	metal–organic chemical vapor deposition
MOCZ	manganese oxide–coated zeolite
MW	microwave
PCE	power conversion efficiency
SSA	specific surface area
TEM	transmission electron microscopy
TGA	thermal gravimetric analysis
XRD	X-ray diffraction techniques, X-ray crystallography

Nomenclature

a	lattice constant (Å)
c	concentration
C	concentration (mg/L), (mmol/L)
C_0	initial metal ion concentration (mg/L), (mmol/L)
c_{BET}	BET constant
C_e	equilibrium metal ion concentration in (mg/L), (mmol/L)
$E\%$	extraction efficiency in percentage
E_g	bandgap
H	magnetic field strength (kA/m)
K_L	equilibrium adsorption constant (L/mmol)
m	weight of substance in grams
M	magnetization of the particles (A·m^2/kg)
n	amount absorbed
n_m	monolayer capacity
P	pressure (Pa)
P_0	initial pressure (Pa)
q	sorption capacity (mg/g), (mmol/g)
q_e	adsorption equilibrium metal ion uptake capacity (mmol/g)
q_{max}	maximum amount/weight of metal ion per unit weight of a sorbent (mg/g), (mmol/g)
R	reflectance
R^2	correlation factor of model approximation

s		scattering coefficient
SSA		specific surface area (m²/g)
T		infrared transmission (%)
t		temperature (°C)
v		water-to-surfactant ratio (–)
V		volume
η		percentage yield (%)
τ		time

Introduction

This book has been focused on the application of common nanomaterials for the removal of metallic species that are found in aqueous environment. The "most common" metallic species are As, Cd, Cr, Cu, Pb, Hg, and Ni. Sb, Pd, Pt, U, and Th are the other miscellaneous metallic species where nanoparticles are applied for their removal from aqueous solutions and water (Dubey et al. 2017). It is important to develop robust, eco-friendly, and economically viable treatment methods for the removal of heavy metals from the aquatic system (Vardhan et al. 2019).

Zinc sulfide finds its applications in optics, electronics, laser technologies, solar energy solutions, etc. Zinc sulfide, with addition of small amounts of suitable activator, is used as phosphor in many areas (Bower et al. 2002). For example, when silver is added, the emitted color is bright blue, whereas manganese yields in orange-red color and copper provides greenish color. Electroluminescent materials based on ZnS perform high brightness and are used as flat vacuum-free light sources in panels, scoreboards, or screens for fluoroscopy (Mateleshko et al. 2004). Zinc sulfide is also used as an infrared optical material, planar or shaped into a lens (Gupta and Gupta 2016). Zinc sulfide–based scintillators have the largest light output per event in the family of imaging scintillators used so far in fast neutron radiography (Wu et al. 2013). Apart from widely known optical properties, ZnS nanocrystals were reported to serve as a sorbent for Cu(II) removal from water (Xu et al. 2016).

Despite a broad interest, general synthetic methods to prepare such materials are lacking, due to at least two reasons. First, it is a significant challenge to ensure homogenous and atomic-scale dispersion of the dopant within the matrix. Second, it is difficult to prepare powders, films, or nanoscale wires made out of zinc oxides and sulfides. Moreover, most synthetic methods require stringent processing conditions or very high temperatures (Acton 2013).

Additional issue is concerning green technologies. Intense development of the industry, power engineering, and agriculture continuously increases pollution of the air (Guttinkunda et al. 2019), water (Xie et al. 2020), and soil (Baltas et al. 2020), which have negative impact on human health. Despite the numerous proposed methods of contamination removal, such as filtration, sedimentation, reverse osmosis, chemical deposition, biological treatment, and so on, sorption methods remain one of the most economical effective ones (Jaspal and Malviya 2020). That is why, sorption is often applied in the final stages of wastewater treatment. A variety of the sorption materials are used, both natural, such as clays, minerals, plant materials, or biomaterials, and synthetic ones. The main advantage of the natural sorption materials is relatively low cost, whereas synthetic ones perform much higher sorption capacity. The aforementioned challenges pose some limitations on the production of the sorption materials, which must be greener itself through minimizing the energy consumption and other environment unfriendly factors. Thus, one of the trends in sorbent synthesis leads toward production of the same materials using different methods, with modification of the sorbent characteristics through different synthesis conditions. In this book, to achieve greener technology, the researches were focused on the water solutions without employment of any organic solvents.

This book proposes some solutions that address aforementioned challenges. Theoretical background of the sorption is described in detail in many works (Dąbrowski and Tertykh 1996; Ronco and Winchester 2001; Dragan 2014), but only experimental research makes it possible to evaluate practical advantages of the synthesized sorbent.

In this book, experimental data on the synthesis of micro- and nanoparticles of zinc, copper, and cadmium sulfides; iron oxides; and manganese oxyhydroxide are systematized and discussed. Much attention is paid to the processes of sulfide particles formation out of thiourea solutions, as well as to the thermal decomposition of precursors. The effect of decomposition processes on the morphological and structural properties of the obtained micro- and nanoparticles is emphasized. Especially emphasized are the sorption characteristics of the particles and their dependence on the synthesis conditions.

Who Should Read This Book?

The book is devoted to the researchers, students, and specialists who are interested in the inorganic synthesis and properties of the sorption materials. The presented material can be helpful as a review of methods and as a handbook of experimental research. It requires at least basic knowledge on chemistry and physics.

How This Book Is Organized?

In this book, we investigate the problems of the controlled synthesis of inorganic compounds and effect of their morphological characteristics on their sorption capacity. As the research objects, sulfides of divalent metals (zinc, cadmium, and copper), iron oxides, and manganese oxyhydroxide were chosen.

This book consists of introduction, two chapters, final remarks, and references.

In the first chapter, controlled synthesis of zinc, cadmium, and copper sulfide particles and their sorption properties are discussed. First, main methods of metal sulfides synthesis are described, emphasizing the impact of synthesis conditions on the structural and morphological characteristics of the obtained particles. After the discussion, investigations were focused on the method of particle precipitation from thiourea solutions. In this respect, detailed results for particle formation of zinc, copper, and cadmium sulfides in various synthesis conditions, such as pH, anionic composition, temperature, and microwave activation, are presented and discussed. And finally,

Introduction

the sorption characteristics of the synthesized particles are described in connection with particles size and morphology and synthesis conditions.

The second chapter is dedicated to the production of iron oxides (hematite and magnetite) and manganese oxyhydroxide particles. Experimental data concerning the effects of iron oxides α-Fe_2O_3, γ-Fe_2O_3, and Fe_3O_4 particles synthesis conditions on their structural and morphological characteristics. Next, the issues of the particles formation of manganese oxyhydroxide are discussed. Then, the experimental sorption characteristics of the synthesized particles are presented.

At the end of monography, final remarks and references are added.

How to Use This Book?

The main objective of this book is to provide possibly full review of the results related to the topic. This book will be interesting to anyone who works in the field of inorganic synthesis or is interested in particles of various inorganic compounds synthesis methods.

In this book, the experimental research results are presented in full. They may be helpful in any work aimed to control structural and morphological characteristics of sulfides of zinc, copper, and cadmium; iron oxides; and manganese oxyhydroxide, especially ones focused on obtaining certain sorption capacities. Moreover, the proposed approach may be applied to other compounds as well.

Acknowledgments

This chapter is made Open Access through funding by Precision Machine Parts Poland Sp. z o.o.

1

Synthesis of Zinc, Copper, Cadmium, and Iron Sulfides and Their Sorption Properties

1.1 Methodology of Synthesis

One of the most important steps in the study on metal sulfides sorption properties was the analysis of their synthesis methods. Different methods provide different particle sizes and surface morphology; thus, it was crucial to choose the proper method. The analysis was performed for zinc sulfide as an example.

1.1.1 Overview and Classification of Synthesis Methods

Due to its wide applications and potential use in high technology field, ZnS was one of the most studied luminescence, photocatalyst, special morphology, and mesoporous materials in inorganic synthesis (Wang M. et al. 2011). Zinc sulfide nanoparticles can be prepared by different methods, such as colloidal aqueous and micellar solution synthesis method, using ultrasonic waves, microwaves (MWs), and gamma irradiation (Tiquia-Arashiro and Rodrigues 2016). Zinc sulfide can be synthesized in the form of various nanostructures, and Fang et al. (2011) point out three types of them:

- 0D nanostructures (nanocrystals of quantum dots type, core/shell nanocrystals, and hollow nanocrystals),
- 1D nanostructures (nanowires, nanorods, nanotubes, nanobelts, nanoribbons, and nanosheets), and
- 2D nanostructures.

Various surface morphologies of the ZnS particles may be obtained by different synthesis methods. To make the analysis more clear, we propose that those methods may be divided into two main groups:

- high-temperature methods (solid-state reaction or thermal evaporation of zinc powder and sulfur powder, thermal decomposition of precursors, etc.),

- synthesis from aqueous and nonaqueous solutions (solvothermal and hydrothermal methods, precipitation from aqueous solutions, etc.).

The reaction may be activated traditionally by high temperature only or by ultrasonic, or by MWs.

1.1.1.1 High-Temperature Methods

Zinc and sulfur react with each other violently to produce zinc sulfide. To obtain pure ZnS, the reaction should be performed in vacuum to prevent the oxidation processes. After the mixture of zinc and sulfur is heated, the elements react with each other at temperature range between 400°C and 600°C. The reaction is highly exothermic. Temperature at which reaction is initiated is dependent on the dispersity of components, i.e., if the particles are smaller, the reaction starts at lower temperature. The reaction may be also activated in the MW field. For instance, Manoharan et al. (2001) demonstrated the capability of MW-aided process to perform in the atomic-level doping of "activator" ions in the host lattice. Stoichiometric amounts of reagent grade zinc and sulfur powders were placed in a quartz tube, which was then evacuated at 10^{-3} Torr and sealed. The contents of the tube were exposed to MW irradiation in a kitchen MW oven. The exposure time for the complete reaction was 20 minutes. The average size of the obtained crystallites was 1 μm. Dhara et al. (2018) reported also successful synthesis of Mn-doped ZnS nanocrystals by mechanically alloying the stoichiometric mixture of elemental Zn, S, and Mn powders at room temperature under inert atmosphere of Ar using a high-energy planetary ball mill. Different proportions of cubic and hexagonal phases were obtained dependent on milling time varying from 2 to 6 hours, with the crystallite size of cubic phase between ~24 and ~40 nm and crystallite size of hexagonal phase from ~15 to ~30 nm. Since the interaction takes place at the temperatures above 400°C, this method is classified as a high-temperature one.

In the present research, the reaction started after 1–2 minutes of MW irradiation of the stoichiometric mixture of zinc and sulfur. Then the mixture heats up very quickly, in several seconds, to the temperature of 500°C–600°C. X-ray analysis informed that the obtained zinc sulfide crystallites had a structure of sphalerite. Both single-phase nanopowders and multiphase compositions perform high surface activity and agglomeration ability (Gevorkyan et al. 2019). Unlike interaction between zinc and selenium where the sintering of ZnSe particles takes place (Sofronov et al. 2013), no sintering was observed during MW-assisted synthesis of ZnS. As it is seen in Figure 1.1, the spherical agglomerated particles are formed, and their size appears to be from 50 up to 200 nm.

Solid-state synthesis is known as a very simple and cost-effective method that enables to obtain pure zinc sulfide nanoparticles (Jothibas et al. 2017). In our research, the infrared (IR) spectrum of the ZnS powder synthesized in the MW field out of zinc and sulfur did not show the absorption bands, as it is seen in Figure 1.2, which proved its high purity. In the IR spectrum, only

FIGURE 1.1
SEM image of the ZnS particles obtained with MW activation. MW, microwave; SEM, scanning electron microscope.

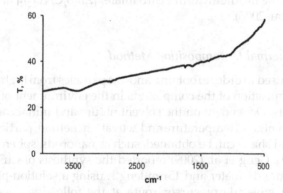

FIGURE 1.2
IR spectrum of the ZnS powder obtained in the MW field. IR, infrared; MW, microwave.

a weak absorption can be observed in the range of 3,200–3,700 cm^{-1}, which was obviously related to the stretching vibrations of water molecules. These molecules were present due to the application of potassium bromide during the IR spectrum analysis.

To reduce the exothermic effect, the well-established technique of chemical vapor deposition (CVD) is applied. CVD methods of zinc sulfide can be separated into static and dynamic methods, as well as transport methods and conventional methods. The standard dynamic ZnS CVD process as practiced today involves a liquid zinc metal that combines with hydrogen sulfide gas in a higher-temperature deposition area (McCloy and Tustison 2013, p. 34). In the CVD method with Zn and S powder as precursors, both substances are loaded into a quartz tube at some distance from one another. In a low-temperature area of the tube, also a substrate or mandrel is placed, so the zinc sulfide can deposit on it. Then the tube is evacuated and heated up to 400°C–800°C, and the carrier gas is injected (usually nitrogen or argon).

Zhuo et al. (2008) reported that with a gas transport method at 650°C, the ZnS nanowires of 50–80 nm diameter and several dozens of microns long were obtained on the aluminum substrate. When the process temperature was increased up to 700°C, the length of nanofibers decreased down to several microns, which turned the nanofibers into a nanorods. The use of graphite substrate caused the forming of nanowires with diameter ca. 50 nm. Similarly, nanowire arrays of cubic zinc sulfide were synthesized on zinc foil by a simple thermal evaporation route at substantially lower temperature, below 500°C (Biswas et al. 2008). Interesting structures of manganese-, copper-, and cobalt-doped ZnS nanotowers are reported to be obtained on Si substrate at 750°C (Zhang et al. 2007).

Zinc sulfide may be obtained also from decomposition of some organic zinc complexes. For example, the polycrystalline ZnS nanotubes with diameters between 140 and 250 nm and the length up to tens of microns were obtained by using metal–organic chemical vapor deposition (MOCVD) template method out of the bis(diethyldithiocarbamate) [$Zn(S_2CNEt_2)_2$] at temperature 400°C (Zhai et al. 2006).

1.1.1.2 Solvothermal Decomposition Method

The method is used in order to obtain nanosize particles from high-temperature thermal decomposition of the compounds in the environment of high boiling organic solvents. Dependent on the solvent nature and initial components, as well as on the synthesis temperature and activation method, particles of various dimensions and shape can be obtained, such as nanorods, spheres, nanowhiskers, and so on. Geng et al. (2009) reported the synthesis of Cu^{2+}-doped ZnS nanorods of 4 nm diameter and 15 nm length using a solution-phase thermal decomposition molecule precursor route at the following conditions: inert environment, oleylamine at 270°C, and reaction time 7 hours. Zhai et al. (2012) reported preparation of rod-, dot-shaped, and quasi-cubic ZnS nanocrystals using oleylamine as coordinating solvent, zinc stearate as Zn precursor and sulfur powder, and thiourea and dodecanethiol as S precursors, respectively. However, when thiourea and dodecanethiol took place of sulfur powder, no nanorods but dot-shaped and quasi-cubic ZnS nanocrystals were formed.

Pure and uniform hexagonal-phase ZnS nanorods with diameters 7–30 nm were synthesized by solvothermal decomposition of zinc diethyldithiocarbamate, $Zn-(DDTC)_2$, in hydrazine hydrate aqueous solutions at 150°C–200°C during 3–48 hours (Zhang et al. 2006). During decomposition of zinc dithiocarbamate in ethanol at 150°C–200°C, hexagonal-phase particles of spherical ZnS are formed, and their dimensions are 30–60 nm (Zhang et al. 2005). Others reported (Abdullah et al. 2016) that the precursor was decomposed to hexagonal zinc sulfide after 2–6 hours of calcination duration at 400°C, and the sizes of zinc sulfide nanoparticles were about 6–11 nm. The existence of the hexagonal ZnS phase was not affected by the calcination duration. Sun et al. (2008) performed MW decomposition of zinc diethyldithiocarbamate at 110°C obtaining the ZnS particles size of 5 nm. ZnS particles of two shapes were reported to be obtained from thiourea solutions at a thermal heating (70°C–90°C) and MW irradiation (Sofronov, Sofronova et al., 2013). The dimensions of agglomerated particles of the spherical shape were 50–100 nm (particle size in agglomerates of 3–5 nm), whereas the length of hexagonal columns was up to 2 μm and diameter 80–100 nm.

Li Y. et al. (2008) proposed a microemulsion-assisted solvothermal process to obtain ZnS nanospheres with rough surface of cubic modification with lattice constant $a = 5.414$ Å. Zinc acetate ($Zn(CH_3COO)_2 \cdot 2H_2O$) was used as starting material, and thiourea (($NH_2)_2CS$) as the source of S, whereas cyclohexane served as the continuous oil phase, cetyltrimethylammonium bromide (CTAB) as the surfactant, 1-pentanol as the cosurfactant. The process lasted for 16 hours at temperature 120°C. Theoretically, diameter of the nanoparticles is directly proportional to the size of the microemulsion water pools, which can be expressed by the water-to-surfactant ratio ($v = [H_2O]/[surfactant]$). When the v value changed from 8, to 16, 24, and 32, the average diameter of ZnS nanosphere samples changed from 200, to 250, 350, and 1500 nm, respectively. When the temperature rose up to 180°C, the diameters of nanospheres with rough surface increased up to 1.25 and even to 3 μm.

When zinc sulfate is contracting with $Na_2S_2O_3$ in cyclohexane with CTAB at 160°C during 12 hours, nanowires of zinc sulfide were obtained of cubic crystal system. The length of the particles was several micrometers, and diameter was 30–50 nm (Dong et al. 2007). ZnS microflowers composed of numerous nanowires with a cubic structure were reported to be synthesized in water below 100°C using an inorganic capping agent $K_2S_2O_7$ (Jaffres et al. 2017). Aguilar et al. (2017) proposed a synthesis of cubic-phase ZnS using synthesized *hydrozincite* $Zn_5(CO_3)_2(OH)_6$ as a novel zinc precursor in hydro- or solvothermal method.

Wurtzite structure of ZnS nanowires was obtained when zinc nitrate contracted with thiourea within the polyethylene glycol (PEG) gel and water at 172°C during 4 days (Cheng et al. 2006). The Ag_2S-catalyzed zinc sulfide nanowires were reported to have a wurtzite structure with a width in the range of 30–80 nm and length of ca. 1 μm (Kim et al., 2018).

Unlike in the case of solvothermal synthesis, in hydrothermal method, water is applied as a solvent. The process is carried out in an autoclave and takes place at increased pressure and temperature. Wang et al. (2017) reported preparation of ZnS nanoparticles by hydrothermal method using zinc nitrate [$Zn(NO_3)_2 \cdot 6H_2O$] and thiourea [$SC(NH_2)_2$] as sources of Zn^{2+} and S^{2-} ions and CTAB [$CH_3(CH_2)_{15}N^+ (CH_3)_3 \cdot Br^-$] as a surface active agent. Hydrothermal synthesis may be initiated by MWs as well (Yan et al. 2013). Anand et al. (2009) proposed optimization of hydrothermal method for the synthesis of ZnS nanoparticles. The nanoparticles were stabilized using hexamethylenetetramine (HMTA) as a surfactant in aqueous solution, and average grain size of the nanoparticles of the order of 2 nm was obtained. Chanu et al. (2017) reported synthesis of zinc sulfide (ZnS) nanospheres using amino acid, L-histidine as a capping agent by hydrothermal method. A particle size of 5 nm was obtained after 3 hours at the temperature 120°C. Other paper reported production of the spherical ZnS nanoparticles with diameter 100–300 nm from the reaction between zinc chloride and $Na_2S_2O_3$ at temperature 200°C during 4 hours (Jiang et al. 2007). Addition of acrylic amide C_3H_5NO caused the increase of the particle diameters up to 3 μm. When the process was performed in the conditions of MW activation, the monodisperse spherical particles of sphalerite structure (a = 5.406 Å) were obtained with diameter ca. 100 nm (Yao et al. 2008). It was also proposed to apply thioacetamide (Zhu et al. 2001), thiourea (Chen et al. 2003), and sodium sulfide (Li W. et al. 2008) as reagents in the hydrothermal synthesis.

Precipitation from a water solution appears to be advantageous because of the simplicity of method and its high productivity. For example, it was reported that pure and Ni (0.5%–2.0%)-doped ZnS nanoparticles were prepared by an inexpensive solid-state reaction method (Jothibas et al. 2018). A solution method for preparing oxygen-doped ZnS colloidal nanocrystals was demonstrated by Wang et al. (2018), where optimal power conversion efficiency (PCE) of the device based on ZnO/ZnS(O) reached 8.85%. The average size of the nanoparticles was ca. 8.5 nm. Song et al. (2008) reported that pure as well as Mn- and Cu-doped zinc sulfide was obtained from the reaction between zinc ethylenediaminetetraacetate and thioacetamide at 60°C–100°C. The dimensions of particles varied between 50 and 1000 nm, dependent on the synthesis temperature and thioacetamide concentration. The smallest particles were obtained at the temperature 60°C and molar ratio TAA:Zn^{2+} = 1:1.

The Mn-doped particles of ZnS (5%–20% in mass) with dimensions 2.5–4 nm were obtained by Wang et al. (2008) in the Ar atmosphere during 30 minutes. Xiao and Xiao (2008) proved that when the thioglycolic acid (TGA) is added, the particles size will be decreased down to 1.2–2.5 nm. Labiadh et al. (2017) reported that undoped and Mn-doped ZnS nanoparticles (with 5%, 10%, and 20% of Mn dopant) were synthesized at 95°C in basic aqueous solution using the nucleation doping strategy. Tripathi et al. (2007) reported synthesis of Mn-doped ZnS nanocrystals via aqueous solution precipitation method with hydrogen sulfide at room temperature in the air atmosphere during 5 hours. Obtained $Zn_{1-x}Mn_xS$ (x = 0%–5%) nanoparticles of cubic modification had sizes in the range of 4–6 nm.

Mei et al. (2017) proposed a facile MW-assisted aqueous route using sodium citrate and TGA as dual stabilizer in order to synthesize Cu–In–S/ZnS quantum dots with the size of 3.8 nm. Zinc sulfide (ZnS), various concentrations of Cu^{2+} (0.25%–1.25%)-doped ZnS, and ZnS:Cu^{2+} nanoparticles capped with various surfactants were reported to be successfully synthesized by a chemical precipitation method in ambient air at 80°C (Murugadoss 2013). The formation of a cubic phase was confirmed for all samples, and average size of the particles ranged from 3.2 to 5.3 nm.

Nanoparticles ZnS(Cu) were obtained from thiourea solution with additions of triethanolamine (TEA) and sodium polyphosphate (Kim et al. 2006; Lee et al. 2004). In other papers, precipitation of ZnS(Cu) was proposed using sodium sulfide (Corrado et al. 2009; Bol et al. 2002). Similarly, precipitation with sodium sulfide enabled to obtain nanoparticles of ZnS(1%Ag) using 3-mercaptopropionic acid (3-MPA) as a stabilizer (Jian et al. 2006).

In this chapter, the methods of metal sulfides synthesis were chosen on the basis of following requirements and assumptions:

- simple technology,
- avoidance of complicated and specific apparatus for synthesis,
- high productivity,
- potential for further increase of the process scale in order to produce larger amounts of the material.

After the thorough analysis, it was assumed that the method best meeting those requirements and providing metal sulfides for further utilization as sorbents was precipitation of them out of the aqueous solutions. As reagents, thiourea and sodium sulfide were chosen.

1.1.2 Methodology and Equipment

MW-activated synthesis was performed using an MW apparatus MARS (CEM Corporation Matthews, USA). The system uses integrated sensor technology to recognize the vessel type and the sample number in order to apply the appropriate amount of power. It has contactless in situ temperature measurement, is equipped with sensors that detect vessel type and count, and provides hundreds of preprogrammed methods.

The specific surface area (SSA) was determined with the method of thermal desorption of argon (10% argon mixture with helium) with chromatographic analysis. As a standard, the samples of aluminum oxide were used with the SSA of 4.2 and 52 m^2/g. The weighed sample (0.05–0.1 g) was placed in a U-tube with inner diameter 4 mm and heated during an hour in the stream of argon of flow speed of 100 cm^3/min at 100°C during 1 hour. This operation served for cleaning the surface and pores from the contaminations and moisture. After that, the gas was replaced by the mixture of 10% argon in helium, and flow speed was reduced down to 50 cm^3/min. The gas mixture is then passed through a katharometer. The tube with the sample was placed in liquid nitrogen where it was until the end of adsorption process indicated by the return of katharometer to the initial (zero) point. It lasts ca. 2–3 minutes. Next, the tube with the sample was placed in water at room temperature where desorption of argon takes place, and the katharometer's indication went to opposite direction. The area under desorption peak is proportional to the overall amount of desorbed argon. It is assumed that the argon adsorption at 77 K has a monolayer characteristics; thus, the area under peak is assumed to be proportional to the absolute surface of the sample. Along with each sample, the standard was measured, i.e., exact weighed sample of Al_2O_3 with known SSA. Two respective U-tubes with measured sample and the standard were connected in series.

Adsorption experiments were performed using the model solutions containing ions of investigated metals at the temperature 20°C (±1°C). Measurement of the metals concentration in the solutions before and after sorption was performed with the inductively coupled plasma atomic emission spectroscopy (ICP-AES) (Moore 1989). The device was iCAP 6300 Duo made by Thermo Fisher Scientific.

The iCAP 6300 uses a simple pressure-controlled nebulizer gas flow or an optional factory-fitted nebulizer gas mass flow controller. Duo version of the instrument enables both radial and axial view. The device applies spectrometer of Echelle type, 52.91 grooves/mm ruled grating, 383 mm effective focal length, and 9.5° UV-fused silica cross-dispersion prism. The wavelength range is 166–847 nm, spectral bandpass is 7 pm at 200 nm, and the device uses high-performance CID86 chip detector. Plasma gas flow is fixed 12 L/min argon, and nebulizer gas is under pressure control from 0 to 0.4 MPa. Auxiliary gas has four fixed flows: 0, 0.5, 1.0, and 1.5 L/min.

The IR spectra were obtained in tablets of KBr with Spectrum One FT-IR Spectrometer (made by PerkinElmer, USA). The device employs improved Michelson interferometer, self-compensating for dynamic alignment changes due to tilt and shear, incorporating high reflectivity first-surface aluminum-coated optics. The detectors are electrically, temperature-stabilized fast recovery deuterated triglycine sulfate (FR-DTGS) or lithium tantalate (LiTaO$_3$). Wavelength range is 7,800–350 cm^{-1} with KBr beam splitter, wavelength accuracy 0.1 cm^{-1} at 1,600 cm^{-1}, and resolution is 0.5 to 64 cm^{-1}. For the FR-DTGS detector, signal-to-noise ratio for KBr optics is 30,000/1 rms, 6,000/1 p-p for a 5-second measurement and 100,000/1 rms, 20,000/1 p-p for a 1-minute measurement. For the LiTaO$_3$ detector, signal-to-noise ratio for KBr optics is 7,500/1 rms, 1,500/1 p-p for a 5-second measurement and 26,000/1 rms, 5,000/1 p-p for a 1-minute measurement.

As for nontransparent powder sample, it is difficult to obtain the absorption spectrum, the diffuse reflection spectrum was used to calculate the bandgap of materials (Kaihara and Sato 2000). It should be remembered that the diffuse reflection spectrum always contains both absorption and reflection spectral components (Nishikida et al. 1995). The relationship of the intensity of the reflected radiation to the concentration is usually presented as follows (Byrn et al. 2017):

$$F(R) = \frac{(1-R)^2}{2R} = \frac{k}{s} = \frac{Ac}{s} \qquad (1.1)$$

where R is reflectance, k is the absorption coefficient, s is the scattering coefficient, c is the concentration of the absorbing species, and A is the absorbance.

Based on the obtained spectra, the bandgap E_g for the examined material was assessed (Katrunov et al. 2010).

Structural characterization of materials was performed with X-ray powder diffraction technique using Siemens D500 X-ray Diffractometer (XRD), with Ni-filtered Cu sources and graphite-diffracted beam monochrometers. In this work, the diffraction patterns were recorded in the 2θ angle range between 10° and 110° at room temperature, with steps of 0.02° and counting time of 10 s at each point.

Particle shape and size are two concepts that are intrinsically connected (Pons and Dodds 2015). Thus, to evaluate morphology of obtained powders, scanning electron microscope (SEM) and transmission electron microscope (TEM) were applied.

The SEM was JSM-6390LV made by Jeol Ltd. (Japan) with a high resolution of 4.0 nm. The JSM-6390 specimen chamber can accommodate a specimen of up to 150 mm in diameter. Standard automated features include autofocus/autostigmator, autogun (saturation, bias, and alignment), and automatic contrast and brightness. Magnification is ×8 to ×300,000 (at 11 kV or higher) and ×5 to ×300,000 (at 10 kV or lower). The TEM was device of ПЭМ-125 type made by SELMI (Ukraine).

The results of measurement and analysis are presented and discussed in the respective chapters, for each presented method.

1.1.3 Procedures and Reactions

The solutions were prepared using the following precursors: $Cu(NO_3)_2 \cdot 3H_2O$, $CuSO_4 \cdot 5H_2O$, $CuCl_2 \cdot 3H_2O$, $Cd(NO_3)_2 \cdot 4H_2O$, $CdSO_4 \cdot H_2O$, $CdCl_2 \cdot H_2O$, $Zn(NO_3)_2 \cdot 6H_2O$, $ZnSO_4 \cdot 7H_2O$, $ZnCl_2$, thiourea, and aqueous solution of ammonia. All the reagents were qualified as chemically pure. All solutions were made using the distillated water.

Precipitation of the zinc, copper(II), and cadmium was performed from 0.1 and 1 M solutions of nitrate, chloride, and sulfate of the respective metal. To the volume of 1 mL of the metal salt solution, the aqueous solution of ammonia was added, or 0.1 M solution of sodium hydroxide until the require pH was reached. The value of pH was varying between 8 and 12. Next, thiourea (Th) was added in proportion of $c(Th):c(Me^{2+})$ 1:1, 2:1, or 4:1. The mixture was then heated up to boiling temperature and boiled during 1 hour. After synthesis, the obtained precipitate was filtered out, washed with distillated water, and dried at the room temperature during 24 hours.

MW-activated synthesis was performed using a MW apparatus MARS (CEM Corporation, Matthews, USA). Volume 50 mL of 0.1 M solution of zinc, cadmium, or copper(II) nitrate, chloride, or sulfate of basicity pH = 8, 10, and 12 was placed in a 250 mL glass. Then the thiourea was added in molar proportion Me^{2+}/Th 1:1, 1:2, or 1:4 with continuous stirring. Next, the mixture was placed in the viala of volume 100 mL and underwent MW activation during 30 minutes at temperature 100°C and 150°C. After the synthesis was

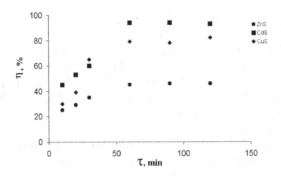

FIGURE 1.3
Percentage of the synthesized sulfides versus precipitation time from thiourea solution for the proportion $c(Me^{2+}):c((NH_2)_2CS)$ 1:1, pH 11.5.

finished, the obtained precipitate was filtered out, several times washed with distilled water, and dried at the room temperature during 24 hours.

The synthesis time was selected based on the assumption of maximal amount of the final product. Figure 1.3 presents the experimental graph of precipitation time for metal sulfides precipitated with thiourea in proportion $c(Me^{2+}):c((NH_2)_2CS)$ 1:1 and pH 11.5.

The experimental data proved that the precipitation time between 10 and 30 minutes provides small amounts of the final products. When increased up to 60 minutes, the time becomes sufficient for obtaining 45%, 79%, and 94% of zinc, cadmium, and copper(II) sulfides, respectively. Further prolongation of the synthesis time has no substantial impact on the product amount. Thus, it was assumed that the basic synthesis time using thiourea solutions was 1 hour.

Iron sulfide was obtained from the thermal reaction between sulfur and iron. Stoichiometric quantities of those substances were placed in the quartz ampule, which was then evacuated with a vacuum pump during 30–40 minutes until the pressure reached 1.3 Pa (10^{-2} mmHg). Next, the ampule was sealed and placed in the oven to be heated up to 600°C at speed of 30°C per hour. The ampule stayed in the temperature 600°C for 5 hours, and then the temperature was risen to 800°C. After 4 hours in the temperature of 800°C, the ampule was cooled down to the room temperature and opened, and the obtained substance was milled in a porcelain mortar.

Acknowledgment

This chapter is made Open Access through funding by Precision Machine Parts Poland Sp. z o.o.

1.2 Precipitation with Sodium Sulfide

Interaction of the metal salts with sodium sulfide can be described as follows:

$$Zn(NO_3)_2 + Na_2S \rightarrow ZnS \downarrow + 2NaNO_3 \quad (1.2)$$

$$CdCl_2 + Na_2S \rightarrow CdS \downarrow + 2NaCl \quad (1.3)$$

$$CuSO_4 + Na_2S \rightarrow CuS \downarrow + Na_2SO_4 \quad (1.4)$$

These reactions are typically applied to obtain metal sulfides insoluble in aqueous solutions, mainly copper(I and II), nickel, zinc, cadmium, lead(II), cobalt, silver, or gold. The precursors in these cases are metal salts soluble in water: halides, sulfates, or nitrates.

Figure 1.4 presents SEM images of the zinc, cadmium, and copper sulfides particles formed during the precipitation process performed with sodium sulfide. It can be seen that in case of zinc and cadmium sulfides, large shapeless agglomerates were formed of dimensions few dozens of micrometers, while in case of copper sulfide, thin scales are formed with dimensions up to 3 μm. However, the analysis of the precipitation conditions impact on the morphological features of the particles revealed no substantial relation between sodium sulfide precipitation process characteristics and the shape or dimensions of the obtained particles. The relative yield of zinc, cadmium, and copper sulfides was ca. 90%–95% which is considered an

FIGURE 1.4
Images of the particles ZnS (a), CdS (b), and CuS (c), obtained by the sodium sulfide precipitation out of 0.1 M solutions.

excellent percentage yield. According to the recognized sources (Furniss et al. 1989, p. 33), the success of a reagent is judged as excellent when the percentage yield is >90%, very good for >80%, good for >70%, fair for >50%, and poor when it is below 40%.

Variations of precipitation temperature and the initial concentration of the substances in the solutions did not effect on morphological features of the obtained particles. In all cases, shapeless large agglomerates were formed. Thus, the method of precipitation with sodium sulfide does not allow to control the formation process and growth of the metal sulfide particles.

1.3 Particles Precipitation Out of Thiourea Solutions

In the aqueous solutions, thiourea undergoes hydrolytic degradation dependent on pH. When the environment is acidic, the ammonium ions NH_4^+ and thiocyanate ions SCN^- appear (Herrmann et al. 1995). The hydrolysis of thiourea in alkaline environment, in turn, leads to the formation of cyanamide and sulfide ions (Peters and Rauter 1974; Kitayev et al. 1989), according to the following reaction:

$$(NH_2)_2 CS + 2OH^- \leftrightarrow S^{2-} + H_2CN_2 + H_2O \quad (1.5)$$

When the alkaline hydrolysis is performed at increased temperatures above 80°C or 100°C, formation of S^{2-}, CO_3^{2-}, and NH_3 takes place. It was assumed that higher temperature caused hydrolysis of cyanamide that resulted with formation of dicyanamide, guanidine, and guanidine-thiourea. These substances, in turn, became hydrolyzed to ammonium cyanate and then to ammonium carbonate (Marcotrigiamo et al. 1972).

Unlike the sodium sulfide precipitation process, the conditions such as pH, concentration, anionic composition, or activation method have substantial effect on the particles formation in thiourea solutions. Thus, it is possible to control in some extent the functional characteristics of the obtained particles varying the synthesis conditions, as it was reported in case of ZnS (Jayalakshmi and Rao 2006; Sofronov, Kamneva et al., 2013).

1.3.1 Effect of Chemical Composition of the Reagent Environment on the Particle Formation of Zinc, Copper, and Cadmium Sulfides

1.3.1.1 pH effect

Figure 1.5 presents powders precipitated from nitrate solutions in relation $c(Zn^{2+}):c((NH_2)_2CS)$ 1:1. Fine bulk powder is formed at pH 8, and the obtained particles are mainly spherical. When pH is increased, diameters of those spherical particles increase, too. For example, at pH 8, the particles of ZnS

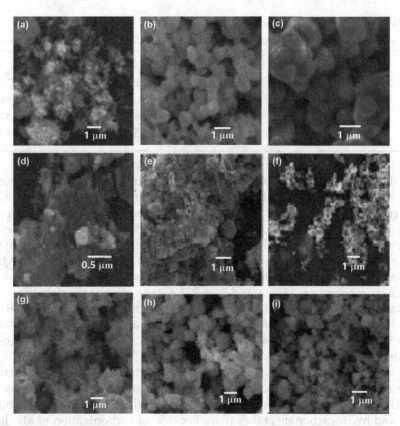

FIGURE 1.5
Particles ZnS (a, b, c), CuS (e, f, g), and CdS (h, i, j), obtained from nitrate solutions at the ratio $c(Zn^{2+}):c((NH_2)_2CS)$ 1:1 and different pH: pH 8 (a, e, h), pH 10 (b, f, i), and pH 12 (c, g, j).

with diameters 0.1–0.2 µm are formed, whereas at pH 10 and 12, the particle dimensions are 0.5 and 0.65 µm, respectively. It can be seen in Figure 1.5a–c. In case of copper sulfide, the particles below 0.1 µm are formed at pH 8, but when pH is increased up to pH 12, the particles grow up to 0.3 µm, as it is shown in Figure 1.5d–f. Similar pattern takes place for the cadmium sulfide, where spherical particles of diameter 0.2 µm are formed at pH 8. When alkalinity is increased up to pH 12, apart from small particles of diameter 0.1–0.3 µm, some larger spherical particles are found of diameter in average 0.75 µm.

Similarly, pH of the solution had an effect on the final product yield. In general, it was observed that the percentage yield can be increased for 5%–15% with the increase of pH value. Table 1.1 provides the examples of obtained yields dependent on pH.

The results of X-ray structural analysis show that from thiourea solutions, zinc sulfide of sphalerite structure (some weak reflexes were noted, too, typical for zinc oxide) and copper sulfide of covellite structure are formed.

TABLE 1.1
pH Solution Effect on the Percentage Yield of the ZnS, CdS, and CuS Precipitation

	Percentage Yield (%)		
pH	ZnS	CdS	CuS
8	40	70	75
10	42	75	77
12	45	85	80

A slight difference takes place in the cadmium sulfide precipitation process (Sofronov et al. 2011). In that case, at the beginning, lemon yellow precipitate is formed, which gradually becomes red when the time of synthesis is prolonged. The first stage, when yellow precipitate is formed, lasts 15 minutes at temperature 90°C–95°C and is longer at lower temperatures. X-ray analysis revealed that the result of synthesis was cadmium sulfide with the following characteristics: the lemon yellow powder was a wurtzite modification with lattice parameters $a = 4.136$ Å and $c = 6.713$ Å, very close to the data reported by Rantala (1999), whereas the red powder was a mixture consisting mainly of sphalerite with lattice constant $a = 5.818$ Å, but also wurtzite modification with lattice parameters $a = 4.136$ Å and $c = 6.713$ Å. Hexagonal wurtzite component in the red powder is no more than 10%.

Apart from zinc sulfide, some amounts of hydroxide, carbonate, hydrocarbonates, and oxides may be produced during the synthesis process, which leads to the contamination of the final product. Formation of carbonates and hydrocarbonates takes place because of carbonization of alkaline solution with the carbon dioxide from atmosphere, whereas zinc oxide is formed by the decomposition of zinc hydroxide. The simplest and most informative method for the inspection of oxygen-containing impurities is the IR spectrometry. The obtained sulfides do not absorb in the near middle IR region, i.e., there are no absorption bands in the spectrum 7800–400 cm^{-1}, while oxygen-containing additions do absorb in their specific spectrum regions.

IR spectra of the synthesized samples are presented in Figures 1.6–1.8. In the IR spectrum of the ZnS powder obtained at pH 8 and proportion $c(Zn^{2+}):c((NH_2)_2CS)$ 1:1 (Fig.ure1.6a), absorption in the region 3,000–3,600 cm^{-1} is seen with its maximum at 34,00 cm^{-1}, as well as the absorption band at 1,620 cm^{-1}. These data correspond with valence and deformation vibrations of water molecules absorbed on the surface of zinc sulfide particles. The absorption band 2,038 cm^{-1} can be ascribed to the vibrations of SCN (Egorov et al. 2010). The absorption band 1,385 cm^{-1} can be ascribed to the vibrations of ion NO_3^- (Adler and Kerr, 1965; Saha and Podder 2011), and the absorption band 900 cm^{-1} can be ascribed to the vibrations of the Zn–OH bond (Adler and Kerr 1965). The absorption band at 1,028 cm^{-1} is related to the symmetric vibrations of C–S bond in thiourea (Saha and Podder 2011).

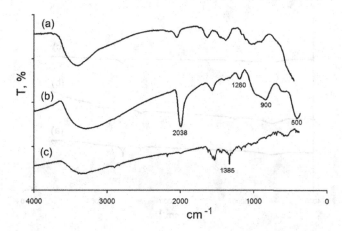

FIGURE 1.6
IR spectra of ZnS powders precipitated from nitrate solutions at various pH: (a) pH 8; (b) pH 10; and (c) pH 12 (Sofronov, Kamneva et al. 2013). IR, infrared.

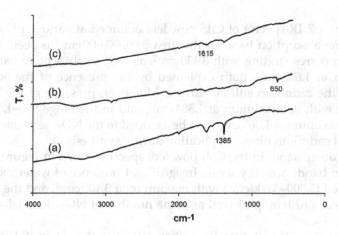

FIGURE 1.7
IR spectra of CdS powders obtained at various pH: (a) pH 8.5; (b) pH 9.5; and (c) pH 12.0. IR, infrared.

When pH is increased up to 10.9, the bands 2038 and 900 cm^{-1} become stronger. Also the bands 1,260, 685, and 500 cm^{-1} are stronger, perhaps because of vibrations of S–O (685 cm^{-1}) and Zn–O (500 cm^{-1}) bonds, as it was reported by Adler and Kerr (1965). The absorption band at 1,260 cm^{-1}, in turn, is generated by thiourea adsorbed on the surface of zinc sulfide particles and corresponds with vibrations of N–CS–N group (Edrah 2010).

However, when alkalinity is increased to pH 12, practically all absorption bands become weaker, as it is seen in Figure 1.6c. This phenomenon proves the decrease of impurities percentage in the obtained sulfide powder.

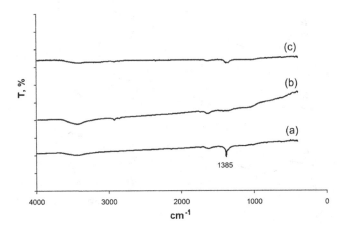

FIGURE 1.8
IR spectra of CuS powders obtained at various pH: (a) pH 8.2; (b) pH 9.5; and (c) pH 12.0. IR, infrared.

In Figure 1.7, IR spectra of CdS powders obtained at various pH are presented. The absorption band in the area 3,000–3,600 cm^{-1} is seen, with its maximum corresponding with 3,430 cm^{-1}, as well as absorption band with maximum at 1,632 cm^{-1}, both explained by the presence of the adsorbed water on the cadmium sulfide surface. Moreover, presence of the absorption band with its maximum at 1,384 cm^{-1}, and in the range 900–1,200 cm^{-1} with its maximum at 1,060 cm^{-1}, can be ascribed to the NO_3^- ions that appear because of cadmium nitrate application during synthesis.

On the other hand, in the CuS powders spectra shown in Figure 1.8, the absorption bands are very weak. Insignificant presence of water molecules was noticed (3,000–3,600 cm^{-1} with maximum at 3,400 cm^{-1}, and the absorption band at 1,620 cm^{-1}), as well as some number of NO_3^- ions (absorption band at 1,385 cm^{-1}).

Investigation of the synthesis conditions effect on the zinc sulfide particles formation leads to the interesting observation. When the aqueous ammonia solution serving as an alkaline reagent was replaced by sodium hydroxide or potassium hydroxide, diameter of spherical particles substantially decreased. That fact is very important for the further investigation of the particle dimensions impact on fundamental absorption edge in cadmium sulfide, especially in case of optical materials production. In order to examine the aforementioned effect, chloride solutions were chosen, because they produced the particles with minimal dimensions and maximal yield.

Figures 1.9 and 1.10 present TEM images of the synthesized zinc sulfide particles. When NaOH was used at low pH 8.1 and low concentration of thiourea in the proportion of $c(Zn^{2+}):c(Th)$ 1:4, the average dimensions of the agglomerates were ca. 20 nm, as it is seen in Figure 1.9a.

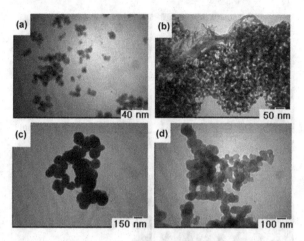

FIGURE 1.9
The particles obtained from chloride solutions (NaOH), at respective pH and proportion $c(Zn^{2+}):c(Th)$: (a) 8.1 and 1:4; (b) 10.9 and 1:1; (c) 10.9 and 1:4; and (d) 12.0 and 1:1 (Sofronov et al. 2014).

When the thiourea concentration and pH increase, the obtained spherical particles become larger. At the proportion $c(Zn^{2+}):c(Th)$ 1:4 and pH 10.9, the average particle dimension is 150 nm (Figure 1.9c), whereas at $c(Zn^{2+}):c(TM)$ 1:1 and pH 12, it is ca. 90 nm (Figure 1.9d). The particles became larger, but in different degree, compared with the ones obtained at pH 8.1.

In case of ammonia hydroxide application as a pH precursor, formation of the spherical agglomerate particles of average diameter 60 nm takes place at pH 8.9 (Figure 1.10a). Here too, when the thiourea concentration and pH increase, the obtained spherical particles become larger with dimensions up to 90 and 150 nm.

It is known that the size of particles can affect the functional features of a material (mechanical, optical, fluorescent, etc.). In the researches, it was found that zinc sulfide synthesis conditions might have effect on the shift of the basic adsorption band edge in the spectra of diffuse reflection (Katrunov et al. 2010; Sofronov et al. 2014). Figure 1.11 presents the diffuse reflection spectra for the obtained powders.

The results of bandgap E_g assessment, synthesis conditions, and average dimensions of the particles are presented in Figure 1.12. The analysis of those data leads to conclusion that the samples obtained with NaOH as a precursor revealed some displacement of the fundamental absorption edge in the range of 20–30 nm dependent on the synthesis conditions. On the other hand, the samples obtained with NH_4OH as a precursor revealed that position of the fundamental absorption edge varied insignificantly.

A wide range of obtained bandgap values versus particle dimensions are seen in Figure 1.12 indicating a variety of quantum size effects. It should be noted that the proposed methodology provides large possibilities to control these parameters through the synthesis conditions, compared with

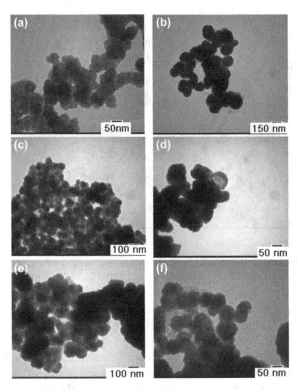

FIGURE 1.10
The particles obtained from chloride solutions (aqueous ammonia solution) at respective pH and proportion $c(Zn^{2+}):c(Th)$: (a) 8.9 and 1:4; (b) 10.9 and 1:1; (c) 10.9 and 1:4; (d) 12.0 and 1:1; (e) 12 and 1:4; and (f) 12 and 1:10 (Sofronov et al., 2014).

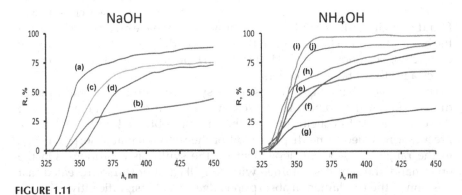

FIGURE 1.11
Diffuse reflection spectra of the ZnS powders precipitated from chloride solutions at pH values regulated by NaOH addition and concentration $c(Zn^{2+}):c(Th)$, respectively: (a) 10.9 and 1:1; (b) 12.0 and 1:1; (c) 8.1 and 1:4; (d) 10.9 and 1:4; (e) 12.0 and 1:1; (f) 8.9 and 1:4; (g) 12.0 and 1:1; (h) 12.0 and 1:4; (i) 12.0 and 1:10; and (j) 12.0 and 1:4 (2 hours) (Sofronov et. al. 2014).

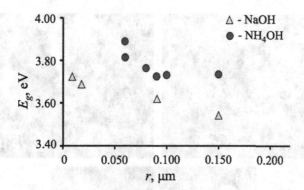

FIGURE 1.12
Dependence of the bandgap E_g on the dimensions of obtained ZnS particles (Sofronov et al. 2014).

the reported values of $E_g = 3.65$ eV for bulk ZnS material and 3.87 eV for MW-assisted self-assembled ZnS nanoballs (Zhao et al. 2004). The graph presents two main correlations:

- For the ZnS powders synthesized with sodium hydroxide, decrease of the particle dimensions caused "linear" shift of the E_g value from 3.54 to 3.72 eV.
- For the ZnS powders synthesized with ammonia hydroxide, in the range of dimensions between 150 and 80 nm, no substantial changes in E_g are seen, but for further dimensional decrease, shift to $E_g = 3.9$ eV takes place.

Important factor influencing these phenomena was the nature of hydroxide applied for the zinc sulfide synthesis, namely, cations Na^+ and NH_4^+. This correlation should be taken into consideration when the initial components are chosen in order to obtain the ZnS powder of desired functional characteristics.

1.3.1.2 Concentration Effect

More concentrated solutions of thiourea doz not provide distinguishable changes of phase composition of sulfides, but they have significant impact on the dimensions of the formed particles. Namely, when the ratio $c(Zn^{2+}):c((NH_2)_2CS)$ increases, dimensions of spherical nanoparticles decrease, as it can be seen in Figure 1.13. Larger thiourea concentrations during the copper(II) sulfide synthesis lead to the formation of spherical agglomerates with following parameters:

- average diameter 0.35 μm (dispersion 0.030) at pH 8 and $c(Cu^{2+}):c((NH_2)_2CS)$ ratio 1:2,

FIGURE 1.13
Images of the ZnS particles obtained from nitrate solutions at respective pH and $c(Zn^{2+}):c((NH_2)_2CS)$ ratios: (a) 8 and 1:4; (b) 12 and 1:4; and (c) 12 and 1:10.

- average diameter 0.54 μm (dispersion 0.022) at pH 8 and $c(Cu^{2+}):c((NH_2)_2CS)$ ratio 1:4,
- average diameter 0.52 μm (dispersion 0.030) at pH 10 and $c(Cu^{2+}):c((NH_2)_2CS)$ ratio 1:2,
- average diameter 0.43 μm (dispersion 0.026) at pH 12 and $c(Cu^{2+}):c((NH_2)_2CS)$ ratio 1:4.

In case of CdS particles, the most significant factor shaping morphological characteristics of the particles is the thiourea concentration in solution. Thus, introduction of the excessive thiourea in two or four times larger amounts causes formation of two types of spherical particles shown in Figure 1.14. Especially in Figure 1.14c, small spherical particles with diameters below 1 μm are clearly seen between the larger formations of thin scales (30–50 nm) grouped together in large balls with diameters ca. 5 μm.

When the molar ratio $c(Me^{2+}):c((NH_2)_2CS)$ is increased, higher yield is obtained, as shown in Figure 1.15. The maximal yield for zinc sulfide is 65%, and that for cadmium and copper sulfides is 95%.

On the other hand, IR spectra of the powders obtained with excessive amounts of thiourea did not reveal any substantial differences. Only the spectra of zinc sulfide powder shown in Figure 1.16 have strengthened

FIGURE 1.14
SEM images of the CdS particles obtained from nitrate solutions at pH 12 and different $c(Zn^{2+}):c((NH_2)_2CS)$ ratios: (a) 1:1; (b) 1:2; and (c) 1:4. SEM, scanning electron microscope.

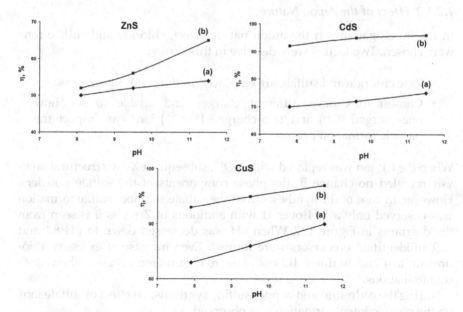

FIGURE 1.15
Graphs of ZnS, CdS, and CuS yield dependent on the pH value at different $c(Me^{2+}):c((NH_2)_2CS)$ concentrations: (a) 1:2 and (b) 1:4.

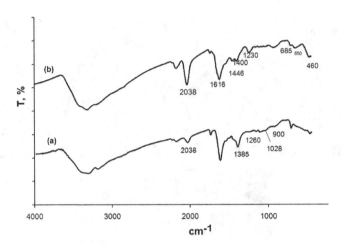

FIGURE 1.16
IR spectra of the ZnS powder precipitated from nitrate solutions at pH 12 and different $c(Me^{2+}):c((NH_2)_2CS)$ concentrations: (a) 1:4 and (b) 1:10. IR, infrared.

absorption band at 2,038 cm^{-1}, caused by SCN vibrations, and absorption band at 1,028 cm^{-1}, which are different from the spectra corresponding with powders obtained at $c(Zn^{2+}):c((NH_2)_2CS)$ ratio 1:1.

1.3.1.3 Effect of the Anion Nature

In the investigations on the anion nature effect, chloride and sulfate ions were chosen. Two factors were decisive in this choice:

- Zinc chloride and sulfate are employed in the synthesis process.
- Chosen ions have different charges and enable to see how one-charged (Cl$^-$) and two-charged (SO_4^{2-}) ions can impact the particle formation process.

When the Cl$^-$ ion was replaced with NO_3^-, subsequent X-ray structural analysis revealed no change in the phase components of the sulfide powders. However, in case of the synthesis from the sulfate solution, sulfite formation was observed only at pH over 11 with additions of ZnO, as it is seen from the diagrams in Figure 1.17. When pH was decreased down to pH 8.1 and 10.9, unidentified precursors were formed. Even increase of excessive thiourea up to 4 and 10 times did not allow to obtain zinc sulfide without ZnO contaminations.

During the cadmium and copper sulfide synthesis, no effect of sulfate ions on the phase content formation was observed.

Images of the obtained zinc sulfide particles out of the chloride solutions are presented in Figure 1.18. It is noteworthy that at pH 8 and $c(Zn^{2+}):c((NH_2)_2CS)$

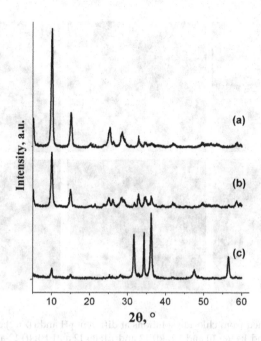

FIGURE 1.17
X-ray analysis of the ZnS powder precipitated from sulfate solutions at different pH and respective $c(Zn^{2+}):c((NH_2)_2CS)$ concentrations: (a) 8.1 and 1:1; (b) 10.9 and 1:1; and (c) 12.0 and 1:1 (Sofronov, Kamneva et al. 2013).

ratio 1:1, the particles were formed as a thick-wall tubes of several microns length (Figure 1.18a). When $c(Zn^{2+}):c((NH_2)_2CS)$ ratio was increased up to 1:4, finely dispersed bulk sediment was formed consisting of particles of various shapes (Figure 1.18b).

When $c(Zn^{2+}):c((NH_2)_2CS)$ ratio was kept 1:1, but pH was increased up to 12, spherical particles 0.5–1.0 µm large were formed, seen in Figure 1.18c and d. However, when pH was kept 12 and $c(Zn^{2+}):c((NH_2)_2CS)$ ratio was increased up to 1:10, the dimensions of spherical particles decreased (Figure 1.18e).

Along with large spherical particles 0.5–1.0 µm, numerous small spherical particles were formed with diameters ca. 0.1–0.2 µm. The average particle diameters obtained at pH 10 and $c(Zn^{2+}):c((NH_2)_2CS)$ ratio 1:1, at pH 12 and $c(Zn^{2+}):c((NH_2)_2CS)$ ratio 1:1, at pH 12 and $c(Zn^{2+}):c((NH_2)_2CS)$ ratio 1:4, and at pH 12 and $c(Zn^{2+}):c((NH_2)_2CS)$ ratio 1:10 were 0.19 µm (dispersity 0.033), 0.54 µm (dispersity 0.034), 0.57 µm (dispersity 0.023), and 0.19 µm (dispersity 0.026), respectively. Hence, during the synthesis from chloride solutions, the smallest spherical particles with diameter 0.1–0.2 µm were formed at pH 12 and $c(Zn^{2+}):c((NH_2)_2CS)$ ratio 1:10.

Figure 1.19 presents SEM images of the particles obtained from sulfide solutions. When $c(Zn^{2+}):c((NH_2)_2CS)$ ratio was increased up to 1:4 and pH up to 11, no significant impact was observed on the morphological features of the particles (Figure 1.19b). However, at pH 12.0 and $c(Zn^{2+}):c((NH_2)_2CS)$ ratios

FIGURE 1.18
ZnS particles obtained from chloride solutions at different pH and c(Zn^{2+}):c((NH_2)$_2$CS) ratios: (a) 8 and 1:1; (b) 8 and 1:4; (c) 10 and 1:1; (d) 12 and 1:1; (e) 12 and 1:4; (f) 12 and 1:10 (Sofronov, Kamneva et al., 2013).

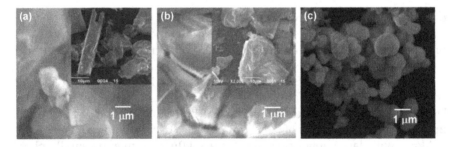

FIGURE 1.19
ZnS particles obtained from sulfate solutions at different pH and c(Zn^{2+}):c((NH_2)$_2$CS) ratios: (a) 8 and 1:1; (b) 8 and 1:4; and (c) 12 and 1:4 (Sofronov, Kamneva et al., 2013).

1:1 and 1:4, the spherical agglomerates are formed along with scales. Further increase of the thiourea concentration leads to the formation of spherical agglomerates with diameters 0.2–1.0 μm only (Figure 1.19c).

It was found that the anion composition change in case of cadmium and copper sulfides precipitation had no significant effect on the phase composition and morphological characteristics of the particles. Usually, finely dispersed sediment is observed, which consists of spherical particles, and their dimensions are larger when the thiourea concentration and pH are increased.

1.3.2 Impact of the Organic Additives

One of the methods to control the morphological characteristics of the nanoparticles is the application of various surface-active additives as a coordinating solvents, in particular, amines, such as ethylenediamine (Mendil et al. 2016; Yao et al. 2005; Behboudnia et al. 2005; Song et al. 2008; Cheng et al. 2006), oleylamine (Geng et al. 2009; Zhai et al. 2012), TEA (Kim et al. 2006; Lee et al. 2004), as well as amino acids, e.g., glycine (Wang et al. 2000; Rakovich et al. 2008). The introduction of valine, proline, serine, cysteine, and methionine allows to control the morphological characteristics of the synthesized cadmium sulfide particles (Qiu et al. 2011).

In the present study, the main criteria for the amino acid choice were two: first, it had to be soluble in water, and secondly, it had to be easily accessible. Thus, the following substances were taken as object for further investigations: glycine, alanine, ε-aminocaproic, glutamic, and p-aminobenzoic acid (Sofronov, Belikov et al. 2013).

SEM images of cadmium sulfide particles synthesized in presence of glycine and alanine are shown in Figures 1.20–1.23. When the synthesis was conducted at the temperature 20°C and glycine content 0.005–0.01 M, large spherical particles resembling flowers were formed with diameters ca. 5 μm, irrespective of thiourea concentration, as it is seen in Figures 1.20a–f. In general, this additive plays no significant role in the cadmium sulfide precipitation process. When the concentration was 0.05 M, morphology of the particles changed dependent on the proportion of cadmium to thiourea C_{Cd}/C_{Th}. When C_{Cd}/C_{Th} was 1:1, the sediment was formed as small spherical particles with diameters ca. 100 nm (Figure 1.20g). When C_{Cd}/C_{Th} was 1:2, small plates were formed (Figure 1.20h), When C_{Cd}/C_{Th} was 1:4, large, dense spherical particles appeared, with dimensions between 10 and 20 μm (Figure 1.20i).

Increase of the concentration up to 0.1 M caused formation of the large shapeless particles seen in Figure 1.20k and l. The higher the thiourea concentration, the larger the particles formed. Addition of the glycine 0.5 M caused formation of the dense spherical particles with diameters ca. 0.5–1.0 μm, irrespective of thiourea concentration, as it is seen in Figure 1.20m.

Unlike the glycine, after addition of 0.005 M alanine, the particles resembling flowers were formed (Figure 1.21), and their dimensions were smaller for larger concentrations of thiourea. When C_{Cd}/C_{Th} was 1:1, the agglomerate dimensions were above 5 μm (Figure 1.21a). When C_{Cd}/C_{Th} was 1:4, dimensions were smaller, ca. 2–3 μm (Figure 1.21b). Increase of alanine concentration caused destruction of "flowers," and formation of the shapeless agglomerated sediments consisted of small plates or scales (Figure 1.21c–k). When amount of alanine was further increased up to 0.5 M, shapeless layered sediments were formed, as it is seen in Figure 1.21l. At the same time, when proportion of cadmium to thiourea C_{Cd}/C_{Th} was increased up to 1:4, dense spherical particles of dimensions ca. 1 μm were formed (Figure 1.21m).

The results were quite different when precipitation was performed at the temperature 100°C. When glycine was added in amount of 0.005–0.010 M,

FIGURE 1.20
CdS particles obtained at 20°C and respective different glycine concentrations and proportions C_{Cd}/C_{Th}: (a) 0.005 M and 1:1; (b) 0.005 M and 1:2; (c) 0.005 M and 1:4; (d) 0.01 M and 1:1; (e) 0.01 M and 1:2; (f) 0.01 M and 1:4; (g) 0.05 M and 1:1; (h) 0.05 M and 1:2; (i) 0.05 M and 1:4; (j) 0.1 M and 1:2; (k) 0.1 M and 1:4; and (l) 0.5 M and 1:2.

finely dispersed sediments were formed, consisting of spherical particles with diameters ca. 1 μm, as it is shown in Figure 1.22a–e. However, an increase of the proportion of cadmium to thiourea C_{Cd}/C_{Th} caused the opposite effect than that observed at room temperature. Namely, at 20°C, thiourea concentration increase caused formation of spherical particles, whereas at 100°C, small scales appeared, especially distinguishable at 0.01 M and C_{Cd}/C_{Th} proportion 1:4 (Figure 1.22e). Further increasing the glycine amount up to 0.05 M caused formation of the particles resembling flowers of dimensions 2–4 μm irrespective of the thiourea concentration (Figure 1.22f–h). Still further increase of glycine concentration up to 0.1–0.5 M caused formation of dense spherical particles of dimensions ca. 0.1–0.3 μm (Figure 1.22i–m).

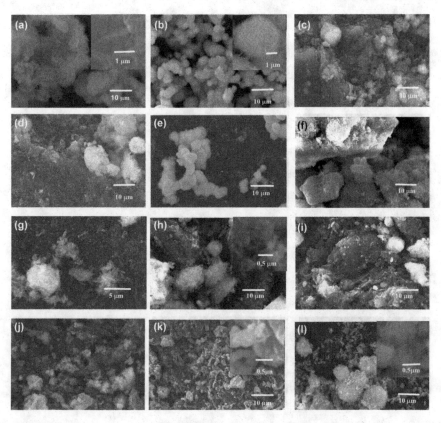

FIGURE 1.21
CdS particles obtained at 20°C and respective different alanine concentrations and proportions C_{Cd}/C_{Th}: (a) 0.005 M and 1:1; (b) 0.005 M and 1:4; (c) 0.01 M and 1:1; (d) 0.01 M and 1:2; (e) 0.01 M and 1:4; (f) 0.05 M and 1:1; (g) 0.05 M and 1:4; (h) 0.1 M and 1:1; (i) 0.1 M and 1:2; (j) 0.1 M and 1:4; (k) 0.5 M and 1:1; and (l) 0.5 M and 1:4 (Sofronov, Belikov et al. 2013).

Similar differences were observed in case of alanine application at 100°C. Then, for C_{Cd}/C_{Th} proportion 1:1 and alanine amount 0.005 M, sediments are much like in case of glycine application (Figure 1.23a), and with larger concentrations of thiourea, the "flowers" became a little more dense (Figure 1.23b). Increasing the amount above 0.01 M causes formation of shapeless agglomerates of single small scales (Figure 1.23c and d). When the alanine amount was increased up to 0.05–0.10 M, flower-like structures were formed, and their dimensions were dependent on the C_{Cd}/C_{Th} proportion. When the alanine concentration was increased up to 0.5 M at C_{Cd}/C_{Th} proportion 1:1, large layered agglomerates were formed (Figure 1.23k), whereas at C_{Cd}/C_{Th} proportion 1:4, the layered morphology changed into flower-like forms (Figure 1.23m).

Thus, it can be summarized that introduction of alanine caused more significant effect on the particle formation, compared with the glycine.

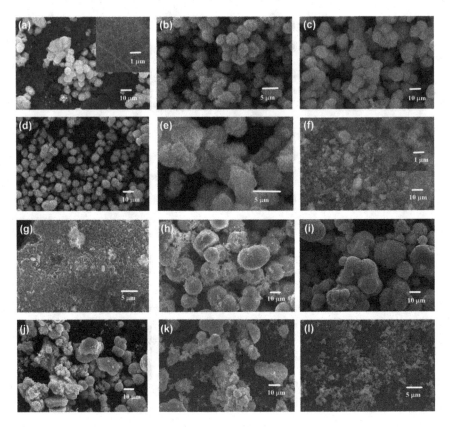

FIGURE 1.22
CdS particles obtained at 100°C and respective different glycine concentrations and C_{Cd}/C_{Th} proportions: (a) 0.005 M and 1:1; (b) 0.005 M and 1:4; (c) 0.01 M and 1:1; (d) 0.01 M and 1:2; (e) 0.01 M and 1:4; (f) 0.05 M and 1:1; (g) 0.05 M and 1:2; (h) 0.05 M and 1:4; (i) 0.1 M and 1:1; (j) 0.1 M and 1:2; (k) 0.1 M and 1:4; and (l) 0.5 M and 1:4.

Images of the cadmium sulfide particles obtained at the presence of ε-aminocaproic acid are shown in Figures 1.24 and 1.25. Addition of 0.005–0.100 M of ε-aminocaproic acid into the synthesis at room temperature causes formation of the agglomerated bilk sediments seen in Figure 1.24a–f. When the amount of ε-aminocaproic acid was increased up to 0.5 M at C_{Cd}/C_{Th} proportion 1:1, formation of the microblocks with dimensions of several micrometers was observed (Figure 1.24g). Higher C_{Cd}/C_{Th} proportion 1:4 caused spherical parts formation of two sorts (Figure 1.24h): smaller ones with dimensions 0.5–1.0 μm and larger ones resembling flowers with dimensions ca. 10 μm. When synthesis was performed at 100°C and ε-aminocaproic acid concentration 0.005–0.100 M, mainly spherical flower-like particles were formed (Figure 1.25a–f) with dimensions between 5 and 10 μm. It can be noted that higher concentrations of thiourea caused formation of more homogeneous particles. When thiourea concentration was increased up to

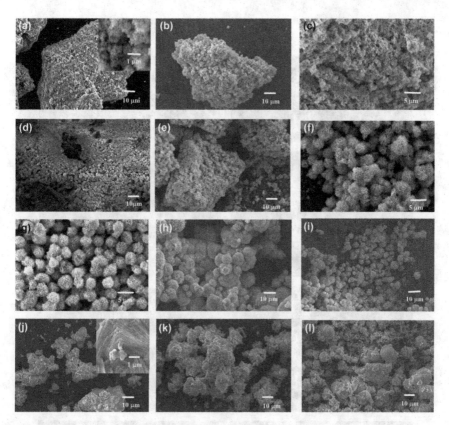

FIGURE 1.23
CdS particles obtained at 100°C and respective different alanine concentrations and C_{Cd}/C_{Th} proportions: (a) 0.005 M and 1:1; (b) 0.005 M and 1:4; (c) 0.01 M and 1:1; (d) 0.01 M and 1:4; (e) 0.05 M and 1:1; (f) 0.05 M and 1:2; (g) 0.05 M and 1:4; (h) 0.1 M and 1:1; (i) 0.1 M and 1:4; (j) 0.5 M and 1:1; (k) 0.5 M and 1:2; and (l) 0.5 M and 1:4 (Sofronov, Belikov et al., 2013).

0.5 M at C_{Cd}/C_{Th} proportion 1:1, spherical agglomerates were destroyed and bulk sediment appeared (Figure 1.25g). Increase of the thiourea concentration made it more dense (Figure 1.25h).

Thus, compared with additions of glycine and alanine, even small amounts of ε-aminocaproic acid (up to 0.1 M) added to the reaction at room temperature caused formation of mainly bulk sediments. Comparative analysis of glycine, alanine, and ε-aminocaproic acid revealed that the longer the carbon chain in the amino acid structure, the stronger its effect on the particles formation. Increase of the synthesis temperature up to 100°C, however, neutralized the action of ε-aminocaproic acid and resulted in formation of large particles resembling flower shape.

Addition of the glutamic acid in concentration range between 0.005 and 0.05 M at 20°C caused formation of the bulk sediments irrespective of the thiourea concentration (Figure 1.26). It was observed that the larger the

FIGURE 1.24
CdS particles obtained at 20°C and respective different ε-aminocaproic acid concentrations and C_{Cd}/C_{Th} proportions: (a) 0.005 M and 1:1; (b) 0.005 M and 1:4; (c) 0.01 M and 1:1; (d) 0.01 M and 1:4; (e) 0.1 M and 1:1; (f) 0.1 M and 1:4; (g) 0.5 M and 1:1; and (h) 0.5 M and 1:4 (Sofronov, Belikov et al. 2013).

glutamic acid concentration and the smaller the C_{Cd}/C_{Th} proportion, the more friable the bulk sediment. Increase of the synthesis temperature up to 100°C caused no considerable change in the particles morphology.

In Figures 1.27 and 1.28, CdS particles obtained at 20°C and 100°C with additions of p-aminobenzoic acid are presented. Solubility of the p-aminobenzoic acid is poor; some sources provide percentage 0.34% at 12.8°C

FIGURE 1.25
CdS particles obtained at 100°C at respective different ε-aminocaproic acid concentrations and C_{Cd}/C_{Th} proportions: (a) 0.005 M and 1:1; (b) 0.005 M and 1:4; (c) 0.05 M and 1:1; (d) 0.05 M and 1:4; (e) 0.1 M and 1:1; (f) 0.1 M and 1:4; (g) 0.5 M and 1:1; and (h) 0.5 M and 1:4.

(Knunyantz 1983), and others inform that 1 gram of p-aminobenzoic acid dissolves in 170 ml water or in 90 ml boiling water (Brittain 1993). Taking this into consideration, experiments were performed only with its concentration 0.005 M. Irrespective of temperature and C_{Cd}/C_{Th} proportion, mainly bulk sediment was formed.

FIGURE 1.26
CdS particles obtained at 20°C and respective different glutamic acid concentrations and C_{Cd}/C_{Th} proportions: (a) 0.005 M and 1:1; (b) 0.005 M and 1:4; (c) 0.01 M and 1:1; (d) 0.01 M and 1:4; (e) 0.05 M and 1:1; and (f) 0.05 M and 1:4 (Sofronov, Belikov et al. 2013).

FIGURE 1.27
CdS particles obtained at 20°C at respective different p-aminobenzoic acid concentrations and C_{Cd}/C_{Th} proportions: (a) 0.005 M and 1:1; (b) 0.005 M and 1:4; and (c) 0.01 M and 1:1.

FIGURE 1.28
CdS particles obtained at 100°C at respective different p-aminobenzoic acid concentrations and C_{Cd}/C_{Th} proportions: (a) 0.005 M and 1:1; (b) 0.005 M and 1:4; and (c) 0.01 M and 1:4.

Thus, addition of glycine in amounts 0.005–0.010 M irrespective of thiourea concentration caused formation of large spherical flower-like particles with dimensions ca. 5 μm. Increase of its amount caused formation of large shapeless particles, and their dimensions were larger for higher concentrations of thiourea. When the synthesis temperature was increased up to 100°C, at glycine amounts below 0.01 M, finely dispersed sediments were formed. The higher the C_{Cd}/C_{Th} ratio, the smaller the obtained particles. At the same time, increase of the additive amount up to 0.05 M caused formation of the flower-like particles with dimensions 2–4 µm, irrespective of thiourea concentration.

On the other hand, when small amounts of alanine were added (0.005 M), the flower-like particles were formed, but their dimensions were smaller when thiourea concentration was higher. Further increase of the alanine concentration caused formation of shapeless agglomerated sediments consisting of small plates or scales. When the synthesis temperature was increased up to 100°C at alanine concentrations below 0.05 M, the shapeless agglomerates were formed consisting of single small scales, whereas increase of the alanine amount up to 0.1 M caused formation of structures resembling flowers, and their dimensions were dependent on the C_{Cd}/C_{Th} proportions.

Even small amounts of ε-aminocaproic acid (0.1 M) added at room temperature caused formation of the bulk sediment. However, increase of the synthesis temperature up to 100°C neutralized effect of the ε-aminocaproic acid. As a result, large flower-like particles were formed.

Addition of the glutamic or p-aminobenzoic acid caused formation of mainly bulk sediment, irrespective of other synthesis conditions.

1.4 Effect of Microwave Activation on the Zinc, Cadmium, and Copper Sulfide Particles Formation

1.4.1 Effect of pH and Thiourea Concentration Combined with Microwave Activation

MW irradiation is widely applied in synthesis of nanoparticles of various applications (Qiao et al. 2017). For example, Zhu et al. (2001) proposed a method of CdS and ZnS nanoparticles preparation through the reaction between $CdCl_2$ or $ZnZ(Ac)_2$ and thioacetamide in aqueous solution using MW irradiation. However, production of zinc sulfide out of thiourea solutions with MW activation was not reported before.

Application of MWs allowed to obtain the individual phases of metal sulfides at pH close to 8. Figure 1.29 presents the X-ray diagrams of zinc sulfide powder, as well as cadmium and copper(II) sulfides powder synthesized with MW activation at $c(Me^{2+}):c((NH_2)_2CS)$ ratio 1:1 and pH 8. In the diagrams, only reflexes related to the metal sulfides are seen. In case of zinc sulfide, sphalerite phase is observed, whereas for copper(II) sulfide, it was CuS covellite, and for cadmium sulfide, the hawleyite structure was formed.

IR spectrometry data also proved the decrease of oxygen-containing impurities concentration in the synthesized substances (Figures 1.30 and 1.31).

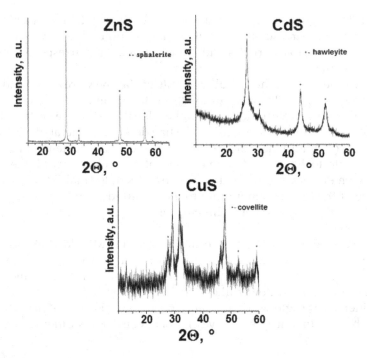

FIGURE 1.29
X-ray diagrams of the powders obtained with microwave activation at $c(Me^{2+}):c((NH_2)_2CS)$ ratio 1:1, pH 8, and temperature 90°C.

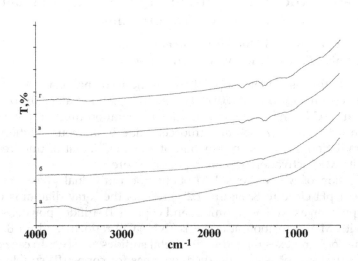

FIGURE 1.30
IR spectra of CdS powder precipitated from the nitrate solutions at temperature 90°C and $c(Me^{2+}):c((NH_2)_2CS)$ ratio 1:1 and different pH values: (a) pH 8; (b) pH 10; (c) pH 11; and (d) pH 12. IR, infrared.

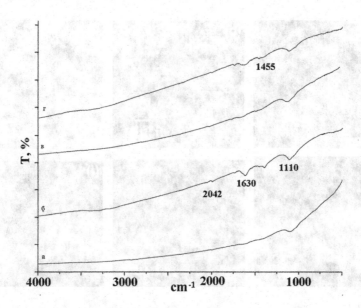

FIGURE 1.31
IR spectra of CuS powder precipitated from the nitrates solutions at temperature 90°C, with different $c(Me^{2+}):c((NH_2)_2CS)$ ratios and respective pH values: (a) 1:1 and pH 8; (b) 1:1 and pH 10; (c) 1:1 and pH 12; and (d) 1:4 and pH 12.

In the spectra of powders obtained with MW activation, compared with IR spectra of powders synthesized with thermal activation, considerable decrease of absorption band intensity in the range of 500–1,500 cm^{-1} is observed. This indicates that impurities concentration is smaller in the obtained substances.

MW irradiation allowed to increase substantially both purity of the obtained compounds and their practical percentage yield. Thus, irrespective of pH and thiourea concentration, yield of ZnS, CdS, and CuS was excellent: 90%–92%, 93%–95%, and 95%–97% by mass, respectively. A slight tendency of yield increase in the series of ZnS, CdS, and CuS is seen, and the thiourea concentration had no substantial effect on the product yield.

SEM images of the ZnS particles are presented in Figure 1.32. In general, synthesis with MW activation did not change the particle forms, and mainly spherical particles were formed. Thus, at pH 8.5 and 9.5, when the $c(Th):c(Zn^{2+})$ proportion was 1:1, diameters of formed particles were below 100 nm with average ca. 80 nm, but single large particles were of diameters ca. 1.5 µm (Figure 1.32a). Further increase of pH up to 10.8 caused increase of particle dimensions up to 0.3–1.5 µm with average value 0.36 µm (Figure 1.32c–d). When sodium hydroxide was applied, some plates of length and width ca. 1.2–1.5 µm and thickness 0.1–0.2 µm were formed along with spherical particles (Figure 2.30e). Increase of pH up to 12 caused formation of spherical particles with diameters 0.2–0.7 µm (Figure 1.32f).

FIGURE 1.32
ZnS particles obtained at temperature 90°C and c(Th):c(Zn^{2+}) proportion 1:1 with different pH: (a) 8.5; (b) 9.5; (c) 10.8; (d) 12; (e) 8 (NaOH); and (f) 12 (NaOH).

In case of cadmium sulfide, unlike zinc sulfide, at low values of pH 8–9, the formation of agglomerates was observed resembling flowers of dimensions 5 μm consisting of plates with thickness ca. 50–80 nm (Figure 1.33a). When pH was increased up to 10, spherical dense particles with diameters 0.2–0.7 μm were formed, along with some flower-like particles with dimensions 0.8–1.0 μm. Further pH increase had no substantial effect on the particles morphology (Figure 1.33b–c). However, when the alkalinity providing reagent was replaced with sodium hydroxide, exclusively spherical particles were formed with average dimensions ca. 0.4 μm (Figure 1.33d–f).

In case of copper sulfide, only spherical particles were formed, and their dimensions were larger with pH increase (Figure 1.34). At pH 8, obtained particle dimensions were below 50 nm, but when pH increased up to 10 and 12, dimensions were larger, up to 200 nm (Figure 1.34a–c). When sodium hydroxide was applied to regulate alkalinity, the agglomerates were formed consisting of spherical particles below 100 nm (Figure 1.34d–f).

Increase of thiourea concentration c(Th):c(Zn^{2+}) in the solution during the synthesis of metal sulfides up to 1:4 and more had varying impact, as it can be seen in Figures 1.35–1.37. In case of zinc sulfide, the particles in agglomerates became larger up to 1–2 μm at pH 8, but increase of pH caused reduction of the dimensions down to 0.6–0.8 μm at pH 12. In case of cadmium sulfide, increase of the thiourea concentration caused formation of the finely dispersed sediment consisting of the particles smaller than 100 nm (Figure 1.36a). When pH increased,

FIGURE 1.33
CdS particles obtained at temperature 90°C from nitrate solutions with $c(Cd^{2+}):c((NH_2)_2CS)$ proportion 1:1 with different pH: (a) 8; (b) 10; (c) 12; (d) 8 (NaOH); (e) 10 (NaOH); and (f) 12 (NaOH).

FIGURE 1.34
CuS particles obtained at temperature 90°C from nitrate solutions with $c(Cu^{2+}):c((NH_2)_2CS)$ proportion 1:1 with different pH: (a) 8; (b) 10; (c) 12; (d) 8 (NaOH); (e) 10 (NaOH); and (f) 12 (NaOH) (Bulgakova et al. 2016a).

FIGURE 1.35
ZnS particles obtained at temperature 100°C and c(Th):c(Zn^{2+}) proportion 1:4 with different pH: (a) 8; (b) 9.5; (c) 10.7; and (d) 12.

FIGURE 1.36
CdS particles obtained at temperature 90°C with c(Cd^{2+}):c($(NH_2)_2CS$) proportion 1:4 with different pH: (a) 8; (b) 10; (c) 12.

particles increased, too, and flower-like formations appeared (Figure 1.36c). In case of copper(II) sulfide precipitation, increased thiourea concentration caused formation of large spherical agglomerates with dimensions up to 1.5 µm. Their dimensions went smaller along with increase of pH (Figure 1.37).

1.4.2 Temperature Effect

When the temperature of synthesis was increased up to 150°C, considerable change in the zinc and copper(II) sulfides particles morphology was observed. Thus, in case of ZnS particles obtained at c($(NH_2)_2CS$):c(Zn^{2+}) ratio 1:1 and alkalinity pH 8, smaller spherical particles had diameter 0.05–0.10 µm and larger

FIGURE 1.37
CuS particles obtained at temperature 90°C from nitrate solutions with $c(Cu^{2+}):c((NH_2)_2CS)$ proportion 1:4 with different pH: (a) 8; (b) 10; and (c) 12.

FIGURE 1.38
ZnS particles obtained at temperature 150°C and $c((NH_2)_2CS):c(Zn^{2+})$ proportion 1:1 with different pH: (a) 8; (b) 9.5; (c) 10.5; and (d) 12.

ones had diameter 0.8–1.5 μm (Figure 1.38). Average diameter of all particles was 0.16 μm. At higher pH 9.5, particles in general grew larger with average diameter 0.44 μm. Further increase of pH up to 10.5 caused smaller content of large particles, and average diameter was 0.29 μm and dispersity was 0.025.

SEM images of the particles CdS obtained from $c(Th):c(Cd^{2+})$ proportion 1:1 at temperature 150°C are presented in Figure 1.39. When alkalinity was pH 8, the geometrically spherical particles were formed along with deformed spheres. Diameters of spherical particles were ca. 0.6–1.0 μm. At higher pH, formed spheres had larger diameters ca. 0.8–1.2 μm.

FIGURE 1.39
CdS particles obtained at temperature 150°C and c(Th):c(Cd^{2+}) proportion 1:1 with different pH: (a) 8 and (b) 12.

FIGURE 1.40
CuS particles obtained at temperature 150°C and c(Cu^{2+}):c((NH$_2$)$_2$CS) proportion 1:1 with different pH: (a) 8; (b) 10; and (c) 12.

As it is seen in Figure 1.40, no substantial impact of higher temperature on the morphological characteristics of particles was observed, unlike in case of ZnS and CdS.

1.5 Effect of Metal Ion Doping on the Zinc Sulfide Particles Formation

There are many published reports on the methodology of doped particles of metal sulfides synthesis (Alby et al. 2018). Recent papers emphasize, on the one hand, problems of green synthesis (Muraleedharan et al. 2015) and, on the other hand, energy-saving synthesis and high efficiency sorptive capture of radionuclides (Vellingiri et al. 2018). It is well documented that copper-doped, manganese-doped, and silver-doped zinc sulfides are excellent phosphors with good photoluminescence properties

(Murugadoss 2013; Vadiraj and Belagali, 2017). Addition of the europium and cerium ions to the particles of ZnS and CdS has considerable impact enabling their photoluminescence properties (Syamchand and Sony 2015; Hurma 2016). However, the effect of the doping ions on the particles formation is still little known.

In order to examine the issue of dopant effect on the particles formation, zinc sulfide was chosen as a basic substance, whereas copper, silver, manganese, europium, and cerium were chosen as dopants. Concentration of dopants was varied between 0.01% and 5% by mass. Synthesis was performed at pH 12 and $c(Zn^{2+}):c((NH_2)_2CS)$ ratio 1:1. The precipitation temperature with thermal activation was 100°C, whereas the one with MW activation was 100°C and 150°C. Duration of synthesis was 1 hour for thermal activation and 30 minutes for MW activation.

It is well known that copper and silver salts react with thiourea producing respective sulfides. In case of powders consisting of ZnS–CuS obtained at thermal activation, no considerable effect of the copper concentration on the particles formation was observed. Always spherical agglomerated particles were formed with diameters between 0.5 and 1.2 μm, as it is seen in Figure 1.41. Similarly, precipitation with MW activation at temperature 100°C did not provide any substantial change in the particles morphology even when the copper concentration was increased up to 1% by mass. Always, the spherical particles with diameters between 0.2 and 0.6 μm were formed, shown in Figure 1.42. However, when Cu concentration was 1% by mass, large plates (scales) with several microns dimensions and thickness 0.2 μm were observed along with spherical particles. When copper concentration

FIGURE 1.41
ZnS–CuS particles obtained at thermal activation with different concentrations of additives (% by mass): (a) 0.01; (b) 0.05; (c) 0.1; (d) 0.5; (e) 1; (f) 5.

FIGURE 1.42
ZnS–CuS particles obtained at microwave activation (100°C) with different concentrations of additives (% by mass): (a) 0.01; (b) 0.05; (c) 0.1; (d) 0.5; (e) 1; (f) 5.

was further increased up to 5% by mass, the particles resembling flowers were formed, as in case of cadmium sulfide.

When the synthesis temperature was increased up to 150°C, spherical particles of diameters between 0.8 and 1.2 μm were formed at the copper concentration 0.01%–1% by mass (Figure 1.43). Higher concentrations, 5% by mass, caused formation of rods several microns long with diameters of 0.3 μm.

Unusual feature was observed when the IR spectra of the obtained copper ions–doped zinc sulfide particles were analyzed. Namely, increase of the copper concentration caused considerable intensity decrease of the gap bands, as it is seen in Figure 1.44e and f, which proved decrease of the contaminations concentration. In general, the analyzed IR spectrum was similar to the spectra obtained for copper sulfide powders, where gap bands are practically absent. It is reasonable to formulate a hypothesis that perhaps in the solution where copper concentration is over 0.05% by mass, the ZnS particles became coated with CuS layer forming core–shell structures similar to the reported ones (Thuy et al. 2014). X-ray phase analysis provided data that the particles consist of different phases of zinc sulfide with sphalerite structure and copper sulfide with covellite structure. In the IR spectra of ZnS–CuS particles obtained at MW activation (150°C), no intensive bandgaps are seen, as in case of the spectra shown in Figure 1.44e and f obtained at copper concentration 0.05% by mass and higher.

As in case of ZnS–CuS particles, synthesis of the ZnS–Ag_2S at different concentrations of silver sulfide between 0.1% and 5% by mass with thermal activation did not reveal any substantial differences. The spherical agglomerated particles were formed with dimensions between 0.5 and

FIGURE 1.43
ZnS–CuS particles obtained at microwave activation (150°C) with different concentrations of additives (% by mass): (a) 0.01; (b) 0.05; (c) 0.1; (d) 0.5; (e) 1; (f) 5.

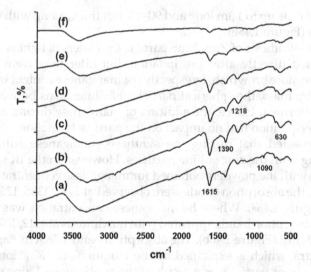

FIGURE 1.44
IR spectra of ZnS–CuS powders obtained at thermal activation with different concentrations of additives (% by mass): (a) 0.01; (b) 0.05; (c) 0.1; (d) 0.5; (e) 1; and (f) 5. IR, infrared.

1.1 μm. However, when MW activation was applied at temperature 100°C, increase of copper sulfide concentration up to 1% by mass caused increase of the spherical particles up to 1 μm. When the synthesis temperature was increased up to 150°C, spherical particles of diameter 0.8–1.2 μm were formed at the Ag_2S concentration in the range of 0.05%–1% by mass (Figure 1.45a–d). Further increase of the silver sulfide concentration up to 5% by mass caused

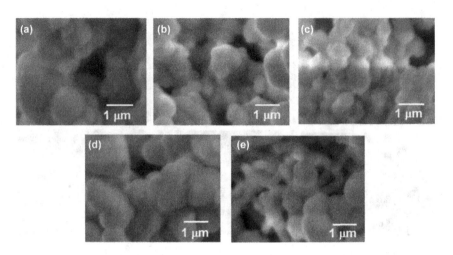

FIGURE 1.45
ZnS–Ag$_2$S particles obtained at microwave activation (150°C) with different concentrations of additives (% by mass): (a) 0.05; (b) 0.1; (c) 0.5; (d) 1; and (e) 5.

formation of rods up to 1 μm long and 90–110 nm thick, along with the spherical particles (Figure 1.45e).

During the synthesis of ZnS–MnS particles, powders of light reddish color were obtained directly after precipitation, but after they were dried, the color became more brownish. Irrespective of manganese content in the range 0.05%–10% by mass, the spherical particles of dimensions between 0.6 and 1.2 μm were formed. Hence, the addition of manganese during zinc sulfide particles precipitation had no impact on the particle formation.

It was expected that during the synthesis, manganese sulfide would appear along with the zinc sulfide particles. However, in the IR spectra corresponding with the powders obtained for manganese concentrations above 1% by mass, the absorption bands were observed at 1442, 1385, 1225, 636, and 460 cm^{-1} (Figure 1.46a). When the manganese concentration was increased, additional absorption bands appeared with maximums at 1112, 1076, 980, 680, 614, and 480 cm^{-1} (Figure 1.46b). The absorption band 1,385 cm^{-1} appeared in all the spectra, which is explained by the vibrations of NO_3^- ion adsorbed on the surfaces of particles as a result of the application of zinc nitrate as a precursor. The absorption bands at 1,442 and 1,076 cm^{-1} corresponded with vibrations of CO_3^{-2} anion, which was formed as a result of alkaline solutions carbonization processes.

Absorption band at 636 cm^{-1} may be ascribed to the vibrations of OH$^-$ groups (Nyquist and Kagel 1971), whereas the band at 460 cm^{-1} may be ascribed to the vibrations of Zn–O bond (Sofronov, Bielikov et al. 2013). Absorption bands at 1225, 1112, and 614 cm^{-1} may correspond with sulfate ion vibrations (Nyquist and Kagel 1971). They are adsorbed on the surface of particles, and

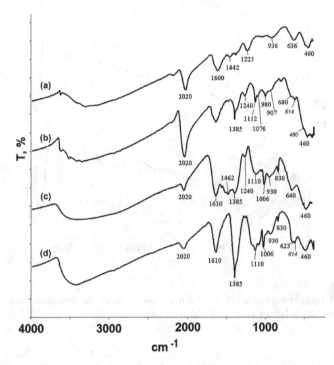

FIGURE 1.46
IR spectra of ZnS particles obtained during precipitation process from thiourea (a, b) and sodium sulfide (c, d) with additions of manganese 1% (a, c) and 10% (b, d) (Sofronov et al. 2017). IR, infrared.

their appearance is a result of manganese sulfate application as a precursor of manganese. Absorption bands at 680, 614, and 480 cm^{-1} appeared in case of the samples with high manganese content and corresponded with the vibrations of Mn–O bond. Presence of the absorption bands with peaks at 614 and 480 cm^{-1} is typical for the vibrations of Mn–O bond in Mn_3O_4 (Baykal et al. 2007). Similar situation took place when the ZnS–MnS system was precipitated with sodium sulfide. Initially, reddish powder was formed, but during drying process, it gradually became brownish. No difference was observed when the synthesis was performed in an inert atmosphere, and subsequently, the powder was dried in vacuum. The brownish color appeared anyway. In the IR spectra of the particles precipitated with sodium sulfide, the same absorption bands are seen as in case of samples obtained from thiourea solutions (Figure 1.46c and d).

In order to confirm Mn_3O_4 formation as a result of manganese application during ZnS synthesis, additional experiments were performed to observe manganese sulfate reaction with thiourea and sodium sulfide. In the IR spectra of the obtained particles shown in Figure 1.47, intensive

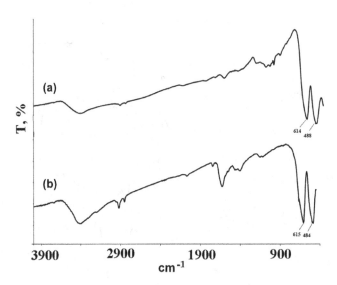

FIGURE 1.47
IR spectra of manganese oxide obtained during precipitation process with thiourea (a) and sodium sulfide (b) (Sofronov et al. 2017). IR, infrared.

absorption bands are seen at 488 and 614 cm^{-1}. X-ray structure analysis provided information that the main phase of the synthesized powders was Mn_3O_4. It was formed as a result of manganese hydroxide and sulfide oxidation with the air oxygen during drying process. Thus, we were unable to obtain the manganese sulfide in the precipitation process from the aqueous solution.

When the copper or silver salts were applied as a dopant during zinc sulfide synthesis, respective sulfides were formed. On the other hand, when the europium or cerium salts were applied, the respective oxides appeared. In case of thermal precipitation of $ZnS-CeO_2$ particles, spherical agglomerated particles were formed. It was found that their dimensions were smaller with the increase of cerium dioxide content. Namely, dimensions of the particles were 0.7–1.1 μm at CeO_2 content 0.05% by mass and 100–200 nm at CeO_2 content 5% by mass (Figure 1.48). MW activation at 100°C also caused decrease of the spherical agglomerates dimensions from 400 down to 100 nm along with increase of dioxide content (Figure 1.49). At higher temperature 150°C with MW activation, large spherical particles of 1–2 μm diameters were formed along with small spherical particles below 0.5 μm (Figure 1.50).

Particles formation process in the system $ZnS-CeO_2$ was similar to the one in the system $ZnS-Eu_2O_3$. In case of both thermal and MW activations, increase of the additive concentration caused decrease of dimensions of spherical agglomerates from 1.2 μm down to 100 nm, as it is seen in

FIGURE 1.48
ZnS–CeO$_2$ particles obtained at thermal activation with different concentrations of additive cerium (% by mass, recounted for CeO$_2$): (a) 0.05; (b) 0.1; (c) 0.5; (d) 1; and (e) 5.

FIGURE 1.49
ZnS–CeO$_2$ particles obtained at microwave activation (100°C) with different concentrations of additive cerium (% by mass, recounted for CeO$_2$): (a) 0.05; (b) 0.1; (c) 0.5; (d) 1; and (e) 5.

Figures 1.51 and 1.52. When the synthesis temperature was as high as 150°C, large spherical particles were formed of dimensions 1–2 µm (Figure 1.53).

Thus, it was found that doping of the zinc sulfide with ions of copper, silver, and manganese had no effect on the particles formation when concentrations were below 1% by mass. However, doping with cerium and europium oxides caused decrease of the particles dimensions from 1 µm down to 100 nm along with increase of the dopant concentrations.

FIGURE 1.50
ZnS–CeO$_2$ particles obtained at microwave activation (150°C) with different concentrations of additive cerium (% by mass, recounted for CeO$_2$): (a) 0.05; (b) 0.1; (c) 0.5; (d) 1; and (e) 5.

FIGURE 1.51
ZnS–Eu$_2$O$_3$ particles obtained at thermal activation with different concentrations of additive europium (% by mass, recounted for Eu$_2$O$_3$): (a) 0.05; (b) 0.1; (c) 0.5; (d) 1; and (e) 5.

Zinc, Copper, Cadmium, and Iron Sulfides

FIGURE 1.52
ZnS–Eu$_2$O$_3$ particles obtained at microwave activation (100°C) with different concentrations of additive europium (% by mass, recounted for Eu$_2$O$_3$): (a) 0.05; (b) 0.1; (c) 0.5; (d) 1; and (e) 5.

FIGURE 1.53
Images of ZnS–Eu$_2$O$_3$ particles obtained at microwave activation (150°C) with different concentrations of additive europium (% by mass, recounted for Eu$_2$O$_3$): (a) 0.05; (b) 0.1; (c) 0.5; (d) 1; and (e) 5.

1.6 Iron Sulfide FeS Formation

In general, methodology of iron salts precipitation with thiourea or sodium sulfide does not succeed in production of iron sulfide nanoparticles. During the drying process, collateral oxidation processes lead to the formation of different phases of iron oxides such as hematite Fe_2O_3 and FeOOH. Hence, in order to obtain iron sulfide particles, the solid-phase interaction of iron and sulfur in vacuum quartz ampoule was performed. The temperature of the process was 700°C, and duration was 8 hours. X-ray fluorescence analysis proved that the iron sulfide was obtained (Figure 1.54). No absorption bands are seen in the IR spectrum shown in Figure 1.55, which indicates high purity of the synthesized product (Figure 1.56).

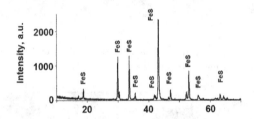

FIGURE 1.54
XRF diagram of the iron sulfide obtained at 700°C solid-phase interaction of iron and sulfur. XRF, X-ray fluorescence.

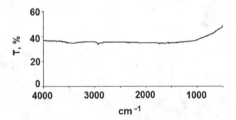

FIGURE 1.55
IR spectrum of the iron sulfide obtained from Fe and S interaction at 700°C. IR, infrared.

FIGURE 1.56
SEM image of the synthesized FeS particles. SEM, scanning electron microscope.

1.7 Sorption Properties of the Metal Sulfides

Sorption is used as one of the technologies for capturing/storing/immobilizing various pollutants and nuclear wastes. Among many materials with different chemical, physical, and structural properties applicable for this purpose (Vellingiri et al. 2018), metal sulfides can be considered.

It was proposed for silver removal to apply copper, cadmium, and nickel sulfides (Cheraneva et al. 2009). The authors demonstrated that in case of applications for dissolution of silver, sorption capacity of sulfides depends not only on the anion compounds but also on the type of salt used for the sulfide synthesis. It was found that the smallest impact on the sorption has nitrate ions. Namely, when the copper, cadmium, and nickel sulfides were extracted out of nitrate solutions and were applied for the silver removal from nitrate solutions, they performed maximal sorption capacity of 437, 365, and 255 mg/g, respectively. However, when those sulfides were applied to remove the silver from solutions AgI_4^{3-} or $Ag(CNS)_2^-$, their capacity reduced dramatically down to 10–30 mg/g.

Barzyk et al. (2002) reported application of the copper sulfide for gold and silver removal. The sorption process here was equivalent to deposition of 50 uniform atomic layers of the metal on the surface of a copper sulfide particle. After sorption, on the copper sulfide surface, some mixed copper and silver sulfides were traced, as well as gold sulfide and crystalline gold. It was noted that the chloride ions have negative impact on the gold extraction process.

Mackinawite's (FeS) ability to adsorp of divalent metals was reported decades ago (Morse and Arakaki 1993). Nanoparticulate mackinawite can be an important host phase for highly effective trapping of mercury (Jeong et al. 2007; Liu et al. 2008; Xiong et al. 2009), lead, and cadmium (Coles et al. 2000). Moreover, iron sulfide is able to capture the ions of arsenic (Niazi and Burton 2016), chromium (Mullet et al. 2004), copper (Watson et al. 2000), nickel (Wilkin and Beak 2017), and zinc (Hamilton-Taylor et al. 1996). Degree of the mercury ions extraction with FeS can reach as high level as 99% at pH 5.6 (Liu et al. 2008). It can be attributed to the oxidation process of the iron sulfide particles in the water solution, with subsequent formation of FeOOH, according to the following reaction:

$$FeS + H_2O + O_2 \rightarrow FeOOH + S^0, \tag{1.6}$$

where the sulfur oxidizes into sulfate ions. 24 hours of aeration leads to the full oxidation of iron sulfide, and then mercury ions adsorption takes place on the surface of FeOOH, which is a good sorbent for Hg. However, sorption capacity of FeS particles for europium is small, ca. 3 mg/g (Allan et al. 2015).

Wershin et al. (1994) examined uranyl cation $[UO_2]^{2+}$ sorption on pyrite (FeS_2). The process is dependent on pH and reaches its highest effectiveness above 98% in the pH range between 4.8 and 5.5.

Perhaps, there is a difference between sorption mechanisms of oxides and hydroxides of metals. For example, it was reported that chromate anion CrO_4^{2-} sorption degree on lead sulfide increases with higher pH values, whereas on hematite (α-Fe_2O_3), it decreases (Musić, 1985). It was reported also that at higher pH, the extraction degree of As(III) ions with iron sulfides FeS and FeS_2 increases. In that case, sorption process is performed according to the following reactions (Bostick and Fendorf 2003):

$$3FeS + As(OH)_3 = FeS_2 + FeAsS + Fe(OH)_3, \qquad (1.7)$$

$$7FeS_2 + 2As(OH)_3 = 3FeS_4 + 2FeAsS + 2Fe(OH)_3. \qquad (1.8)$$

Balsley et al. (1996) did not oppose the hypothesis of similar sorption mechanism on metal sulfides and hydroxides.

For several decades, sorption properties of the metal sulfides are investigated by numerous scholars and teams who pointed out high perspectives of their application (Gong et al. 2016). However, many details concerning stability of the sulfide particles, extraction efficiency, and sorption capacity for heavy metal ions remain still not explored enough (Fu and Wang 2011). In this chapter, the analysis was aimed above all at the sorption properties, such as extraction efficiency and sorption capacity, for metal ions that have radioactive isotopes (such as cobalt, europium, and cerium). The main goal was to describe and evaluate variations of properties in the range of sulfides FeS→ZnS→CdS→CuS, where solubility equilibrium is rising.

1.7.1 pH Impact on the Metal Sulfides Extraction Efficiency and Particles Stability

In the initial stage of the research, the question of the particles stability dependent on pH value was challenged. 1.57 presents the experimental results of metal concentration in solutions after 40 minutes of sorbent treatment. Respective curves correspond with (1) zinc, (2) copper, and (3) iron concentration.

It is clearly seen that at pH below 5, sulfides of zinc, copper, and iron start to dissolve, while at pH higher than 5.5, they are rather stable. Hence, the investigated sulfides can be successfully applied for removal of metals out of water solutions at pH above 5.

Figures 1.58–1.61 present graphs of the metal ions extraction efficiency $E\%$ with the particles of zinc, cadmium, copper, and iron sulfides. The metal sulfides perform their best removal efficiency for europium, cerium, copper, and iron, which reached 100% in the pH range between 5 and 9. When the environment was more acidic, efficiency dropped down, which can be attributed to the dissolution of the sulfides presented in Figure 1.57.

As it is seen from Figures 1.58–1.61, efficiency of zinc, cobalt, manganese, and strontium removal $E\%$ increased for higher pH values of the solutions. The highest efficiency was reached for copper sulfide, whereas the lowest

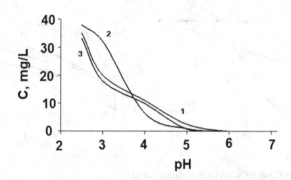

FIGURE 1.57
Graphs of zinc (1), copper (2), and iron (3) concentration in the solutions of different pH.

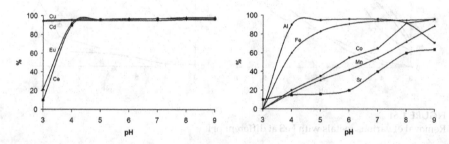

FIGURE 1.58
Removal of various metals with ZnS at different pH.

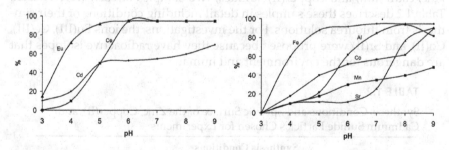

FIGURE 1.59
Removal of various metals with CuS at different pH.

ones for iron sulfide. In particular, the extraction percentage of the copper sulfide for strontium, zinc, cobalt, and manganese at pH 9 reached its maximal value $E\%_{max}$ = 45%, 97%, 98%, and 80%, respectively. In case of the iron sulfide, the extraction efficiency for strontium, zinc, and cobalt at pH reached 35%, 92%, and 70%, respectively.

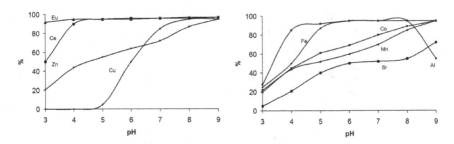

FIGURE 1.60
Removal of various metals with CdS at different pH.

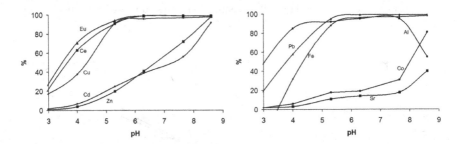

FIGURE 1.61
Removal of various metals with FeS at different pH.

1.7.2 Impact of Specific Surface on the Sorption Capacity of Particles

In order to examine the impact of SSA on the sorption capacity of particles zinc, cadmium, and copper(II) sulfides, two samples of each were chosen. Table 1.2 describes these samples in detail including conditions of their synthesis from thiourea solutions. For the investigations, the ions Eu(III), Ce(III), Co(II), and Sr(II) were proposed because they have radioactive isotopes that are dangerous for the environment and humans.

TABLE 1.2

Synthesis Conditions and Specific Surface of the Zinc, Copper(II), and Cadmium Sulfide Particles Chosen for Experiments

| Sorbent | Synthesis Conditions | | | Specific Surface Area (m²/g) |
	Activation Method	$c(Me2+):c(TM)$	pH	
ZnS-1	Thermal	1:1	10	1.3
ZnS-2	Microwave	1:4	12	30
CdS-1	Thermal	1:1	8	4.2
CdS-2	Thermal	1:1	8	8.5
CuS-1	Thermal	1:1	8	4.5
CuS-2	Thermal	1:4	12	12

The most widely used isotherm equation for modeling equilibrium data is the Langmuir model. The isotherm is valid for monolayer adsorption onto a surface containing a finite number of identical sites. It can be described by the linearized form (Zou et al. 2006):

$$C_e/q_e = 1/(K_L q_{max}) + C_e/q_{max} \tag{1.9}$$

where q_{max} (mmol/g) is the maximum amount of metal ion per unit weight of manganese oxide–coated zeolite (MOCZ), K_L (L/mmol) is the equilibrium adsorption constant, C_e is the equilibrium metal ion concentration in mmol/L, and q_e is the adsorption equilibrium metal ion uptake capacity in mmol/g. By plotting C_e/q_e versus C_e, q_{max} and K_L can be determined.

Figures 1.62–1.64 present sorption isotherms of metal on the zinc, copper, and cadmium sulfides, and the results of their processing are shown in Table 1.3. It should be noted that all the obtained isotherms can be classified as L-type, which is typical for microporous solid bodies with relatively small proportion of the outer surface. The investigated materials performed relatively low sorption capacity (10–20 mg/g), which increases together with the increase of particle SSA. Particles of iron sulfide (obtained with thermal method) for cobalt performed very little sorption capacity (4.5 mg/g). Its sorption isotherm is shown in Figure 1.65.

Thus, it was found that the metal sulfides perform rather small values of sorption capacity, which depend on the SSA of particles.

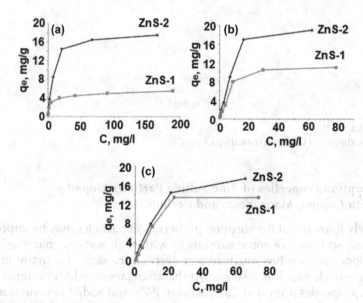

FIGURE 1.62
Sorption isotherms of Co (a), Eu (b), and Ce (c) on the ZnS particles.

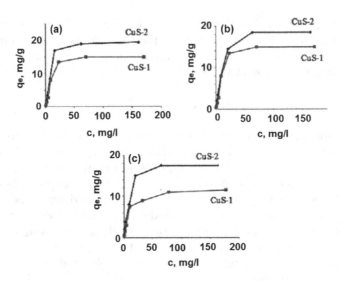

FIGURE 1.63
Sorption isotherms of Co (a), Eu (b), and Ce (c) on the CuS particles.

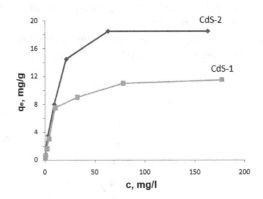

FIGURE 1.64
Sorption isotherms of Co(II) on the CdS particles.

1.7.3 Sorption Properties of Zinc Sulfide Particles Doped with Copper, Manganese, and Cerium

It is widely known that the sorption properties of particles may be improved with small additions of some substances with high sorbent characteristics. Thus, copper sulfide has much higher sorption capacity for many metals than zinc sulfide has. It is also known that manganese oxides perform high sorption properties (Con et al. 2013; Murray 1975) and added to various materials increases their sorption ability. In particular, it was reported that MnO_2-loaded D301 resin was able to remove nearly 100% of cadmium in the range

TABLE 1.3
Parameters of the Langmuir Equation (q_{max} – Sorption Capacity, K_L – Equilibrium Adsorption Constant, R^2 – Correlation Factor) in Case of Sorption of Ce, Co, and Eu on Sorbents ZnS, CdS, and CuS.

Sorbent	SSA (m²/g)	Ce			Eu			Co		
		q_{max} (mg/g)	K_L (L/mg)	R^2 (%)	q_{max} (mg/g)	K_L (L/mg)	R^2 (%)	q_{max} (mg/g)	K_L (L/mg)	R^2 (%)
ZnS-1	1.3	15.5	0.087	98.2	5.6	0.055	97.5	11.2	0.036	98.2
ZnS-2	30	21.4	0.067	98.3	19.3	0.065	96.9	18.9	0.032	91.7
CdS-1	4.2	-	-	-	-	-	-	15.0	0.066	99.9
CdS-2	8.5	-	-	-	-	-	-	21.3	0.072	99.8
CuS-1	4.5	10.1	0.176	99.2	13.7	0.104	99.1	14.2	0.094	97.2
CuS-2	12	18.2	0.115	99.0	18.3	0.094	98.9	23.7	0.065	99.3

SSA, specific surface area.

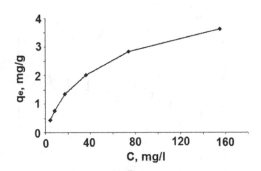

FIGURE 1.65
Sorption isotherm of cobalt on the iron sulfide.

between pH 3 and pH 8 performing sorption capacity for Cd^{2+} ca. 78 mg/g (Zhu et al. 2007).

Another interesting area of investigations is application of manganese oxide–coated zeolite (MOCZ) for the uranium (Han et al. 2007), copper, and lead ions removal (Zou et al. 2006). Sorption isotherms of MOCZ and pH effect on the process were discussed by Han et al. (2007). The authors reported that the equilibrium metal uptake capacity q_e of uranium(VI) was found to be 15.1 mg/g at 293 K and pH 4.0. They calculated it from the formula:

$$q_e = v(C_0 - C_e)/m \tag{1.10}$$

where q_e is the amount of uranium absorbed onto per unit mass of MOCZ at equilibrium in mg/g, v is the sample volume in milliliters, C_0 is the initial metal ion concentration in mg/L, C_e is the equilibrium metal ion concentration in mg/L, and m is the weight of MOCZ in grams.

Interesting result was obtained when the removal capacity of MOCZ was compared with that of raw zeolite. The maximum adsorption capacities for zeolite and MOCZ were found to be 0.061 mmol/g zeolite and 0.108 mmol/g MOCZ for Cu(II), respectively. Adsorption capacities for Pb(II) were 0.134 mmol/g zeolite and 0.243 mmol/g MOCZ, respectively. The results indicated that after introduction of manganese oxide, adsorption capacity of zeolite is almost twice larger than that of raw zeolite for the removal of Cu(II) and Pb(II).

Taffarel and Rubio (2010) investigated MOCZ and its ability in removing Mn^{2+} by adsorption. At pH 6, the attained Mn^{2+} uptake was as high as 1.1 meq Mn^{2+} g^{-1} at equilibrium. Nouh et al. (2015) reported 62% efficiency on the iron adsorption by MOCZ in all studied pH range (0.3–3), and they noted precipitate formed above pH 3, so the efficiency of iron adsorption could not be detected. In three runs, the removal of 93% of the iron content was reached. They reported also that the uranium(VI) stripping efficiency was increased by increasing the aqueous to organic phase ratio to reach 93% at Aq./Org. ratio of 2:1 and reaching maximum (97%) at Aq./Org. ratio of 4:1. Authors calculated the extraction percentage of uranium (VI) from the equation:

$$E\% = \frac{100 D_u}{D_u + V_{aq}/V_{org}}, \quad (1.11)$$

where distribution coefficient D_u was calculated as follows:

$$D_u = \frac{C(\text{org phase})}{C(\text{aq phase})} \times \frac{V(\text{org phase})}{V(\text{aq phase})}, \quad (1.12)$$

where C is the concentration of uranium(VI) and V is the volume.

Chen et al. (2018) applied nanostructured Ce–Mn binary oxide with a Ce/Mn molar ratio of 3:1 to remove arsenite As(III) from water. They obtained maximal sorption capacity of As(III) as high as 97.7 mg/g. The sorbent had an SSA of 157 m²/g with a pore volume of 0.28 cm³/g and a pore size of 7.5 nm, which was higher than that of the pure ceria nanoparticles (86.85 m²/g) and ceria-incorporated manganese oxide (116.96 m²/g).

Based on these reports and experience, particles of zinc sulfide with additions of copper, manganese, and cerium were analyzed to evaluate their sorption capacity for cobalt.

Figure 1.66 presents ion sorption isoterms on the zinc sulfide particles with additions of copper(II) sulfide in proportions 1% and 10%. The graphs can be classified as an L-sort, typical for the solid bodies with micropores, where ratio of outer surface is relatively low. They can be described with the Langmuir model for the cobalt ions adsorption. Obtained values of q_{max}, equilibrium adsorption constant K_L, and correlation ratio R^2 are shown in Table 1.4.

It was found that sorption capacity of the zinc sulfide is almost independent on the amount of added copper(II) sulfide in the range 1%–10% by mass. However, it differs substantially from the sorption capacity of the pure zinc

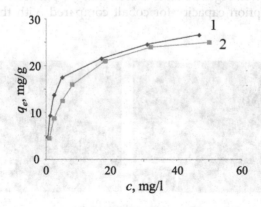

FIGURE 1.66
Sorption isotherm of Co(II) on the iron sulfide with additions of copper sulfide 1% (data 1) and 10% (data 2).

TABLE 1.4

Maximal Sorption q_{max}, Equilibrium Adsorption Constant K_L, and Correlation Ratio R^2 for Ions Co^{2+} Sorption on the Zinc Sulfide with Different Amounts of Copper Sulfide Added

Content of CuS (%)	q_{max} (mg/g)	K_L (L/mg)	R^2 (%)
1	26.1	430	99.9
10	27.1	181	99.8

sulfide which was ca. 19 mg/g for cobalt. Dimensions of the spherical ZnS particles with additions of 1% and 10% CuS were below 100 nm.

In order to examine the effect of manganese presence on the sorption capacity of ZnS particles for cobalt, the zinc sulfide particles doped with manganese 1% (calculated for Mn_3O_4) through precipitation from thiourea solution as well as precipitation with sodium sulfide. In the first case, spherical particles of diameter 0.4–0.7 µm were obtained, while in the last case, their dimensions were below 100 nm. Photomicrographs of the respective particles are shown in Figure 1.67, and cobalt sorption isotherms in Figure 1.68.

When the isothermal data were processed according to the Langmuir model, it was found that sorption capacity for cobalt was 11.2 mg/g in case of particles precipitated with sodium sulfide (dimensions below 100 nm), whereas it was ca. 2 mg/g in case of larger particles 0.4–0.7 µm obtained from thiourea solution. Hence, introduction of manganese did not improve the sorption capacity of zinc sulfide.

To examine the systems of $ZnS-CeO_2$, the particles obtained from thiourea solution at different conditions were used. The respective information on those particles is presented in Table 1.5, whereas sorption isotherms of cobalt are shown in Figure 1.69. It is seen that the particles $ZnS-CeO_2$ perform lower sorption capacity for cobalt compared with that of the pure zinc sulfide.

FIGURE 1.67
Photomicrographs of the particles obtained by (a) precipitation from thiourea solution and (b) precipitation with sodium sulfide.

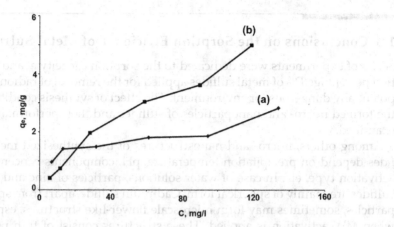

FIGURE 1.68
Sorption isotherms of Co(II) on the zinc sulfide obtained by (a) precipitation from thiourea solution and (b) precipitation with sodium sulfide.

TABLE 1.5
Synthesis Conditions and Sorption Capacity of the System ZnS–CeO$_2$

Sample	Activation Method	Average Particle Dimension (μm)	SSA (m^2/g)
ZnS (0.05 mass% CeO$_2$)	Thermal	220	11.5
ZnS (0.5 mass% CeO$_2$)	Thermal	120	12.0
ZnS (0.5 mass% CeO$_2$)	Microwave	80	30

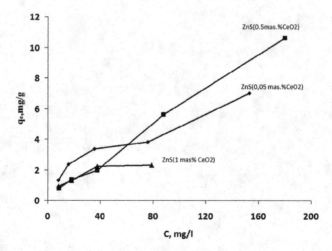

FIGURE 1.69
Sorption isotherms of Co(II) on the particles ZnS–CeO$_2$.

1.8 Conclusions on the Sorption Efficiency of Metal Sulfides

Series of experiments were dedicated to the sorption capacity q_e and extraction percentage $E\%$ of metal sulfides applied for the removal of radionuclides potentially dangerous for environment. The effect of synthesis conditions on the formed micro- and nanoparticles of sulfides and their performance was examined.

Among others, micro- and nanostructures of the synthesized metal sulfides depend on precipitation temperature, pH, components concentration, activation type, etc. In case of water solutions, particles of zinc and copper sulfides are mainly of spherical form. Cadmium sulfide, apart from spherical particles, sometimes may form microscale flower-like structures, especially when MW activation is applied. These structures consist of thin plates of thickness up to 50 nm, joined together into large spherical agglomerates of dimensions above 10 μm. When pH or temperature is increased, as well as when MW activation is applied, the obtained particles are larger. In addition, MW activation causes increased efficiency of the synthesis producing larger amount of metal sulfide. On the other hand, thiourea concentration in the solution leads to the decrease of the particles dimensions.

Metal sulfides are stable at pH above 5. Extraction percentage of the metals on iron, zinc, cadmium, and copper(II) sulfides is higher for higher pH. The most effective performance (above 90%) was noted in case of europium, cerium, copper, and iron uptake at pH above 5. However, they perform rather low sorption capacity for metals such as europium, cerium, and cobalt.

＃ 2

Formation and Sorption Properties of Iron Oxides and Manganese Oxyhydroxide

Among metal oxides, iron oxides are used as sorbents in practical applications (Pepper et al. 2017; Su et al. 2017). Iron oxide nanoparticles (in a variety of chemical and structural forms) have proved their potential in many applications on environmental area (Bhateria and Singh 2019). Their main merits are relatively cheap synthesis and stability of sorption performance. These materials, however, do not act selectively, and their extraction percentage and sorption capacity for metal ions are rather low. The iron oxides perform better sorption characteristics than aluminum oxides.

To increase the effectiveness of iron oxides, a wide range of modifications are being proposed and examined. In case of arsenic removal, Ociński et al. (2014) proposed to use a hybrid polymer containing iron oxides as As(III) and As(V) sorbent for drinking water purification. Moreover, magnetic iron oxides have been proposed to produce adsorbents for natural organic matter, so Lompe et al. (2017) examined effects of iron oxide nanoparticles on the adsorption of organic matter on magnetic powdered activated carbon. Iron oxide nanoflakes were tested in absorption of heavy metals, such as Cd, As, Cr, and Pb (Afridi et al. 2019). Pandi et al. (2017) demonstrated that iron oxide coating on the hydrotalcite/chitosan composite performed good separation ability and displayed an extreme enhanced defluoridation capacity compared with other base components and composites.

Despite the long history of investigations on the sorption properties of iron oxides and their wide practical application, they are still subject of scholarly researches worldwide (Kolida et al. 2014). Cao and Zhu (2008) reported preparation of hierarchically nanostructured α-Fe_2O_3 hollow spheres, whereas Zhu et al. (2013) synthesized hierarchical flower-like α-Fe_2O_3 hollow spheres by one-pot solvothermal method. The latter had high specific surface area (SSA) of $98\,m^2/g$ and a total pore volume of $0.32\,cm^3/g$, which exhibited high removal capacities toward organic dyes and heavy metal ions. For heavy metal ion removal applications, Cao et al. (2012) proposed low-cost synthesis method of flower-like α-Fe_2O_3 nanostructures. Composite sorption materials based on iron oxides were proposed, e.g., by Wang X. et al. (2011) who applied multishelled Co_3O_4–Fe_3O_4 hollow spheres with even magnetic phase distribution for water treatment or by Thirunavukkarasu et al. (2003) who used iron oxide–coated sand for arsenic removal from drinking water. Wu et al. (2005) prepared magnetic powder composite MnO–Fe_2O_3 and examined its ability for the removal of azo dye from water. The iron oxide/carbon nanotubes/chitosan magnetic composite film was proved to be suitable for chromium species removal (Neto et al. 2019).

Another promising oxide is the manganese dioxide, usually applied in the chemical electrical current sources and catalysts, but little known as a sorption material. However, some reports indicate that manganese dioxide performs high sorption properties accompanied by low toxicity and cost-effective synthesis (Nagpal and Kakkar 2019). In particular, SSA of the manganese dioxide is $160\,m^2/g$, and its sorption capacity toward cobalt and zinc at pH 4 is 81 and 65 mg/g, respectively (Loganathan and Burau 1973). Other researchers reported its high sorption capacities toward copper (98 mg/g at pH 5.5), zinc (124 mg/g at pH 6), and lead (80 mg/g at pH 6) (Pretorius and Linder 2001; Dong et al. 2010). Recently, a modified manganese dioxide composite was proposed as an innovative adsorbent for lead(II) ions (Mallakpour and Motirasoul 2019). Our own researches on the sorption characteristics of the manganese dioxides indicated that they perform high sorption properties and can be applied for removal of metal ions from water solutions.

One of the possibilities to control the properties of a material is to form the particles with desired phase composition and morphological characteristics. It is a widely known fact that "the control of composition, size, shape, and morphology of nanomaterials is an essential cornerstone for the development and application of nanomaterials" (Geckeler and Nishide 2010). In other words, controlling dimensions of the particles makes possible to determine their physical and chemical properties (Rasmussen et al. 2018). On the other hand, sorption characteristics of a sorbent depend substantially on its SSA, which, in turn, is determined not only by the particle dimensions but also by porosity and agglomeration degree (Deng et al. 2016). These features are "founded" directly during the synthesis of micro- and nanostructured materials. Accordingly, proper methodology and conditions of synthesis enable to obtain materials with desired properties. Thus, in order to perform controlled synthesis of the final material, it is necessary to know the rules how synthesis conditions effect on the properties of the obtained particles.

Hence, the researches were focused on the relations between the synthesis conditions and formation of micro- and nanoparticles of iron oxide, dioxide, and hydroxide. Subsequent examination of their sorption properties enabled to work out recommendations on the respective sorbent synthesis aimed to obtain improved materials.

2.1 Synthesis of Iron Oxide

2.1.1 Synthesis of the Hematite Particles $\alpha\text{-}Fe_2O_3$

Among many methods to obtain $\alpha\text{-}Fe_2O_3$ particles, two should be emphasized:

- chemical precipitation out of aqueous solution with subsequent high-temperature calcining at 400°C–500°C and
- thermal decomposition of iron salts.

In the first case of chemical precipitation, the hematite particles formation may be influenced not only by the nature of precursors but also by the synthesis conditions, such as pH, temperature, and concentration. Hence, in order to perform controlled synthesis of $\alpha\text{-}Fe_2O_3$ particles with desired morphological features, the effect of those factors should be analyzed.

Figure 2.1 presents scanning electron microscope (SEM) images of the hematite particles obtained through precipitation of iron nitrate with aqueous ammonia solution at various pH. Irrespective of pH value, large agglomerates are formed of random shape, with dimensions between 5 and 200 μm. After high-temperature calcining at 450°C, morphology of the particles did

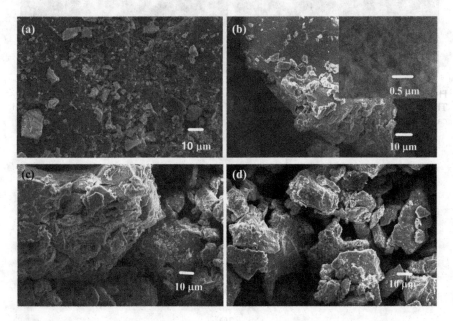

FIGURE 2.1
The particles obtained from nitrate solution at pH 12 (a), pH 9.1 (b), pH 8.3 (c), and pH 7.3 (d).

not change much, as it is seen in Figure 2.2. Precipitation consists mainly of large agglomerates with side dimensions up to 200 μm that are made up of spherical nanoparticles with diameters below 50 nm.

Formation of the iron oxide was confirmed with infrared (IR) spectrometry analysis and with X-ray crystallography. The latter results obtained after high-temperature calcining are shown in Figure 2.3. Irrespective of

FIGURE 2.2
The particles obtained from nitrate solution at pH 12 (a), pH 9.1 (b), pH 8.3 (c), and pH 7.3 (d) and calcined at 450°C for 2 hours.

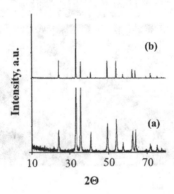

FIGURE 2.3
X-ray diagrams of the powder Fe_2O_3 calcined at 450°C, precipitated from nitrate solutions at pH 12 (a) and pH 7.3 (b).

synthesis conditions, phase α-Fe$_2$O$_3$ (hematite) is formed. In the IR spectra of the powders presented in Figure 2.4, absorption bands are seen in the area of 450–1,000 cm^{-1} at 545 and 555 cm^{-1}. They can be attributed to the vibration of Fe–O bond (Ul-Haq and Haider 2010).

In case of different nature of the iron salt anion, no changes in the particle formation were noted. For instance, when iron chloride was used as a precursor for iron hydroxide particles formation, after subsequent thermal treatment, agglomerates with side dimensions up to 200 μm made up of spherical nanoparticles with diameters below 50 nm were formed, too. The respective SEM image is presented in Figure 2.5.

Thus, it was demonstrated that the prehistory of iron hydroxide synthesis has no effect on subsequent process of the oxide particle formation through calcining.

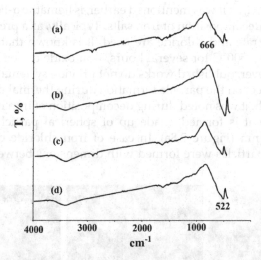

FIGURE 2.4
IR spectra of the powder Fe$_2$O$_3$ calcined at 450°C, precipitated from nitrate solutions at pH 12 (a), pH 9.1 (b), pH 8.3 (c), and pH 7.3 (d). IR, infrared.

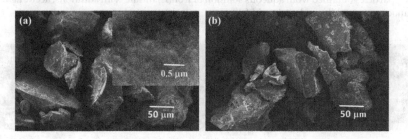

FIGURE 2.5
The iron oxide particles obtained from chloride solution at pH 8.0 (a) and pH 11.5 (b) and calcined at 450°C for 2 hours.

As a rule, large agglomerates are formed with side dimensions from 10 μm up to 200 μm, but with SSA ca. 140 m²/g.

In order to decrease agglomerate dimensions and thus to increase SSA, organic acids were introduced, such as acetic, aminocaproic, and citric ones. They perform complexing properties, and their decomposition during high-temperature calcining may prevent formation of large agglomerates. Figures 2.6 and 2.7 present SEM images of the iron oxide after calcining of the precursors precipitated from aqueous solutions with carbon acids added.

The obtained results indicate that the chosen additions did not have substantial effect on the subsequent formation of the iron oxide during high-temperature calcining. In all cases, large agglomerates were formed with dimensions of hundreds of microns. Probably, the additions had no effect because they were lost during washing procedure before calcining. IR spectra did not reveal any absorption band typical for the added carbon acids.

Another method, as it was mentioned earlier, is hematite α-Fe_2O_3 formation during thermal decomposition of iron salts. Typically, as a precursors in this method, iron nitrate and chloride are used. It is known that after they are calcined at 400°C–500°C for several hours, iron oxide of hematite structure is formed. However, published works do not provide systematic view on the anion nature effect on the particle formation during thermal decomposition.

In our research, it was noted during decomposition of iron nitrate at 450°C that fine sediment is formed, made up of spherical particles with diameters below 100 nm (Figure 2.8a). In case of iron chloride decomposition, oval and cubic particles were formed with dimensions between 1 and 5 μm

FIGURE 2.6
Fe_2O_3 particles obtained from aqueous solution of iron(II) nitrate with additions: acetic acid (a), aminocaproic acid (b), and citric acid (c)

FIGURE 2.7
Fe_2O_3 particles obtained from aqueous solution of iron(III) chloride with additions: acetic acid (a), aminocaproic acid (b), and citric acid (c)

Iron Oxides and Manganese Oxyhydroxide 73

FIGURE 2.8
The particles obtained through decomposition of iron nitrate (a), chloride (b), and fluoride (c) at 450°C during 2 hours

(Figure 2.8b). When iron fluoride underwent calcining, fine sediment was noted made up of spherical and oval particles of dimensions ca. 0.1–0.2 μm (Figure 2.8c). SSA of the particles obtained from iron nitrate was around 140 m²/g, whereas in case of chloride precursor, obtained SSA was no more than 30 m²/g.

Oval particles were formed also when iron nitrate and chloride underwent thermal decomposition together. They are seen in Figure 2.9. It was noted that decrease of chloride compound in the initial mixture caused formation of smaller particles. Presence of the fluoride ions into precursors had no substantial effect on the particles formation.

It was assumed that the fine sediment formation during iron nitrate decomposition is above all influenced by intense release of nitrogen oxide in gas phase. This phenomenon is accompanying the heating process and thus prevents formation of larger iron oxide particles, leaving their dimensions below 100 nm. However, the effect may be caused also by the organic additions.

Figure 2.10 presents the effect of carbon acids additions on the iron oxide particles formation. In all cases, structures made up of submicron particles were formed. However, SSA of the obtained material was varying between 5 and 30 m²/g.

The conclusion is that for the thermal decomposition of salts, the largest SSA of iron oxide particles can be obtained from iron nitrate at 450°C. Obtainable SSA in that method reaches 140 m²/g.

FIGURE 2.9
The particles obtained through decomposition of iron nitrate with addition of $FeCl_3$ in molar proportion 1:1 (a) and 10:1 (b), as well as addition of FeF_3 in molar proportion 1:1 (c).

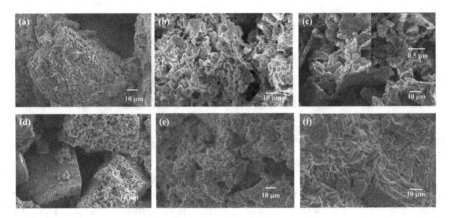

FIGURE 2.10
Fe_2O_3 particles obtained through decomposition of iron nitrate with addition of acetic acid (a, b), aminocaproic acid (c, d), and citric acid (e, f) in molar proportions 1:1 (a, c, e) and 10:1 (b, d, f).

2.1.2 Formation of Maghemite γ-Fe_2O_3 Particles

There were reports on experimental results that indicated similar roles of α-Fe_2O_3 and γ-Fe_2O_3 in removal of elemental mercury (Liu et al. 2015). In order to examine and compare their sorption properties, ferrous oxalate was decomposed at 170°C during 20 minutes. Maghemite formation was confirmed by the X-ray phase analysis from the diagram shown in Figure 2.11. The obtained large agglomerates had side dimensions between 2 and 50 μm

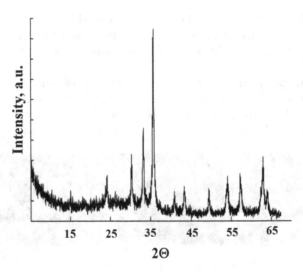

FIGURE 2.11
X-ray diagram of the particles obtained from decomposition of iron(II) oxalate at 170°C.

FIGURE 2.12
The particles obtained from decomposition of iron(II) oxalate at 170°C.

and consisted of faceted particles with side dimensions up to 10 μm, as it is seen in Figure 2.12. SSA of those particles was 35 m²/g.

2.1.3 Peculiarities of the Fe₃O₄ Particle Formation from the Aqueous Solutions

Synthesis of iron oxide nanopowders with high magnetic and sorption properties is for long time of great interest of researchers (Lavrynenko et al. 2018). Cheera et al. (2016) proposed green and cost-effective synthesis of iron oxide Fe₃O₄ magnetic nanoparticles. The obtained nanoparticles were spherical with an average diameter of 20–35 nm, and they exhibited SSA of 26.21 m²/g. Simple and effective method of Fe₃O₄ synthesis is precipitation from aqueous solutions of Fe^{2+} and Fe^{3+} salts at pH above 9 (Odnovolova et al. 2015). However, the particles obtained by this method were reported to perform variations of magnetic properties represented by magnetization – the magnetic moment per weight of the magnetic material – in the wide range from 5 up to 100 A·m²/kg (Baranov and Gubin 2009). No convincing explanation of magnetization variations was given despite many scholars have investigated this phenomenon (Wegmann and Scharr 2018), but some reports suggest their dependence on the particles morphology. For instance, Ge et al. (2009) reported variations of magnetization between 53.3 and 97.4 A·m²/kg attributed to the particles dimensions. In their experiments, iron oxide nanoparticles were synthesized by FeCl₂·4H₂O oxidation in basic aqueous solution at 134°C and pressure 2 atmospheres. Diameters of the particles were tuned from 15 to 31 nm through the variation of the reaction conditions, and their magnetic behavior was either ferromagnetic or superparamagnetic depending on the particle size. Magnetization was noted to be lower for smaller particles. Tombácz et al. (2015) in their review draw the conclusion that in defining the magnetic behavior of iron oxide nanoparticles, the size distribution and morphology are essential. Baaziz et al. (2014) synthesized the iron oxide nanoparticles from iron stearate precursor in the presence of surfactants in high boiling solvents. The average sizes of the particles were obtained in the range 4–28 nm by varying parameters of the thermal decomposition of an iron precursor. It should be highlighted that, according some reports (Roth et al. 2015), formation of particles between 3 and 17 nm with higher saturation magnetization was found directly related to the bigger particle size. These were favored at the highest temperature, the highest iron salt concentrations, a molar ratio of Fe(III)/Fe(II) below 2:1, and a hyperstoichiometric molar ratio of hydroxide ions to iron ions of 1.4:1. Our researches confirmed that temperature of synthesis played a substantial role in the formation of iron oxide magnetic phase.

Figure 2.13 presents X-ray diagrams of the synthesized powders. The samples obtained at 15°C and iron concentration 1 M appeared to be amorphous, as it is seen in Figure 2.13a. Decrease of iron concentration in solution at 15°C did not result with crystalline sediment formation. Only at temperature 20°C, crystalline sediment is formed, but from the solution with iron concentration 1 M, magnetite was not being precipitated (see Figure 2.13b). Formation of pure magnetite was noted at 90°C when concentration of iron in the solution was 0.15 M (Figure 2.13c). At lower temperatures between 20°C and 90°C, X-ray diagrams revealed reflexes that corresponded with magnetite Fe_3O_4, maghemite γ-Fe_2O_3, and goethite α-FeOOH in different proportions.

In Table 2.1, the results are presented on the phase components, as obtained from X-ray diagrams using Rietveld refinement technique. The experimental results indicated that magnetite presence is substantially dependent on the

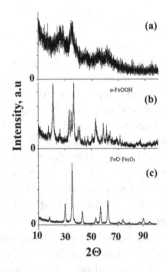

FIGURE 2.13
X-ray diagrams of the particles obtained at 15°C (a) and 20°C (b) from the solution with iron concentration 1 M and obtained at 90°C from the solution with iron concentration 0.15 M (c) (Odnovolova et al. 2015).

TABLE 2.1
Conditions and Results of Fe$_3$O$_4$ Particles Synthesis (Odnovolova et al. 2015)

No.	Synthesis Conditions		Average Particle Diameter (nm)	Obtained Substance	Composition* (wt% (st. deviation))	Average Particle Diameter[a] (nm)
	Iron Overall Concentration (M)	Precipitation Temperature (°C)				
1	0.15	15	9.5	—	—	—
2	0.3	15	9.5	—	—	—
3	1	15	7.0	—	—	—
4	0.15	20		γ-Fe$_2$O$_3$ (maghemite)	7.36 (0.78)	9.9
				Fe$_3$O$_4$-magnetite	85.87 (3.98)	10.8
				α-FeOOH (goethite)	6.77 (0.44)	9.6
5	1	20		α-FeOOH (goethite)	100	8.8
6	0.15	40		γ-Fe$_2$O$_3$ (maghemite)	8.08 (0.05)	8.4
				Fe$_3$O$_4$-magnetite	91.92 (0.79)	10
7	0.15	60	14.6	γ-Fe$_2$O$_3$ (maghemite)	7.29 (0.04)	10.4
				Fe$_3$O$_4$-magnetite	92.71 (0.79)	10.2
8	1	60	10.3	γ-Fe$_2$O$_3$	13.3	7.4
				Fe$_3$O$_4$-magnetite	86.7	8.1
9	0.15	90		Fe$_3$O$_4$-magnetite	100	9.5
10	1	90		γ-Fe$_2$O$_3$ (maghemite)	4.03 (0.52)	7.2
				Fe$_3$O$_4$-magnetite	82.03 (2.14)	7.6
				α-FeOOH (goethite)	13.93 (0.49)	9.4

[a] Obtained using Rietveld refinement technique.

synthesis temperature and iron concentration. Both increase of the temperature and decrease of iron concentration in precursor solution resulted with larger concentration of magnetite phase in synthesized sediment. Eventually, 100% magnetite phase was noted in the samples precipitated at 90°C from solution with iron concentration 0.15 M. On the other hand, at the temperatures below $t = 15°C$, X-ray amorphous oxide was precipitated.

Figure 2.14 presents SEM images of synthesized iron oxide nanoparticles obtained using transmission electron microscopy (TEM). In case of initial iron concentration 0.15 M at temperature 15°C, spherical particles were formed with average diameters ca. 10 nm, as it is seen in Figure 2.14a. When the overall concentration of iron ions was increased up to 0.3 M, it had no substantial effect on the particles' dimensions, which is illustrated in Figure 2.14b. For the powders obtained from 1 M concentration of iron ions, however, dimensions of the particles were distinguishably smaller. Figure 2.14c shows that their average sizes were ca. 7 nm.

On the other hand, higher process temperatures caused growth of the particles. Figures 2.14d and e illustrates the particles with average diameters 15.2 nm obtained from 0.15–0.3 M concentrations, and Figure 2.14f illustrates the particles with average diameters 10.5 nm obtained from solution of 1 M concentration. SSA in all cases was 120–130 m²/g, irrespective of iron ions concentration and synthesis temperature.

Particles' magnetization measurement results are presented in Figure 2.15. Samples obtained at temperature 15°C revealed very low magnetization below 5 A·m²/kg represented by curves a and b. On the other hand, samples

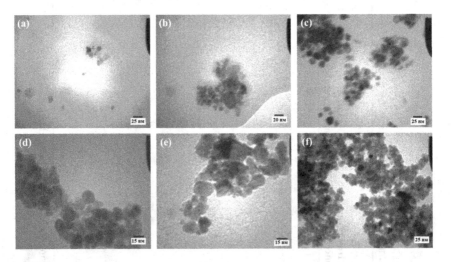

FIGURE 2.14
Fe_3O_4 nanoparticles synthesized from iron chloride and sulfate in $c(Fe^{3+}):c(Fe^{2+})$ proportion 2:1, at pH 8.0–9 and different conditions: (a) $t = 15°C$ and $C(Fe) = 0.15$ M; (b) $t = 15°C$ and $C(Fe) = 0.3$ M; (c) $t = 15°C$ and $C(Fe) = 1$ M; (d) $t = 60°C$ and $C(Fe) = 0.15$ M; (e) $t = 60°C$ and $C(Fe) = 0,3$ M; and (f) $t = 60°C$ and $C(Fe) = 1$ M (Odnovolova et al. 2015).

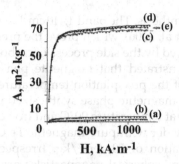

FIGURE 2.15
Magnetization M versus magnetic field strength H for the Fe_3O_4 samples obtained at different temperature t and iron ions concentration: (a) 15°C and 1 M; (b) 15°C and 0.15 M; (c) 60°C and 1 M; (d) 60°C and 0.3 M; and (e) 60°C and 0.15 M (Odnovolova et al. 2015).

obtained at 60°C had high magnetization ca. 70 A·m²/kg represented by curves c and d in Figure 2.15. It is seen that the magnetization is increased a little in case of smaller concentration of iron ions in the initial solution. Obviously, the higher the proportion of magnetite in a sample, the higher the magnetization it reveals. Moreover, higher magnetization can be expected for larger particles in general. Figure 2.16 presents IR spectra of the synthesized iron oxide samples.

It is seen in the IR spectra that there is absorption band in the area 450–1,000 cm⁻¹ with its peak at 572 and 575 cm⁻¹ in the curves a and b, respectively. They are attributed to the vibration of Fe–O bond in the iron oxide Fe_3O_4 (Ai et al. 2008; Odnovolova et al. 2014). Curve b exhibits also absorption bands at 1,128 and 1,410 cm⁻¹, which can be attributed to the presence of SO_4^{2-} ions (Saha and Podder 2011) caused by the application of iron sulfate

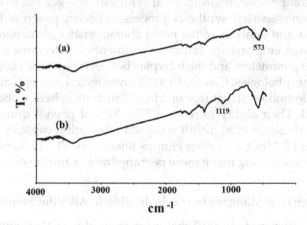

FIGURE 2.16
IR spectra of Fe_3O_4 particles obtained from chloride (a) and sulfate (b) aqueous solutions. IR, infrared.

as a precursor in the synthesis. The band 1,410 cm^{-1} is attributed to the CO_3^{2-} ion vibrations (Hong et al. 2006), which prove the presence of the carbonate contaminations produced by the side process of carbonization.

Hence, it was demonstrated that magnetite phase formation is under substantial influence of the precipitation temperature. The higher the temperature, the more the magnetite phase synthesized in a sample. When the synthesis is performed at temperature 90°C and iron concentration 0.15 M in the initial solution, powder with pure magnetite Fe_3O_4 is formed (100 wt%) that reveals magnetization of 70 A·m^2/kg. Irrespective of the precipitation process parameters, spherical nanoparticles are formed with average dimensions between 7 and 15 nm. Increase of the synthesis temperature and decrease of the iron ions concentration cause formation of the powder made up of larger particles with higher magnetization.

2.2 Synthesis of Manganese Dioxide and Oxyhydroxide

Manganese oxide minerals occur in three polymorphic forms: α-, β-, and δ-MnO_2. Typically, they have large surface area and more passive and active sites, allowing high adsorption and excellent oxidation and catalysis activity (Feng et al. 2007). MnO_2 nanomaterials are the subject of many research works because of their importance and potential technological applications as heterogeneous catalysts for ozone decomposition, organic pollutants oxidation, nitric oxide detraction, carbon monoxide reduction, and degradation of dyes (Ahmed 2016). In his review, the author pointed out a range of synthesis methods such as sonochemical synthesis, solution combustion synthesis, thermal decomposition, hydrothermal synthesis, sol–gel, electrodeposition, and microwave-assisted synthesis processes. Recent progress focused on the synthesis and analysis of the novel characteristics of manganese oxide nanostructures, emphasizing critical experiments to determine the chemical and physical parameters and the interplay between synthetic conditions and nanoscale morphologies. Cao et al. (2009) investigated adsorption properties of microscale hollow structures of Mn_2O_3, such as sphere, cube, ellipsoid, or dumbbell. Their ability to remove 77%–83% of phenol from water was demonstrated. Singh et al. (2010) analyzed adsorption capacity of α-MnO_2 nanorods and δ-MnO_2 nanofiber clumps toward As(V). The nanostructures were synthesized using manganese pentahydrate in an aqueous solution.

2.2.1 Synthesis of Manganese Oxyhydroxide in Alkaline Solutions

It is known (Turner et al. 2008) that interaction of ions Mn^{2+} and MnO_4^- in alkaline environment produces manganese dioxide according to the following reaction:

$$3MnCl_2 + 2KMnO_4 + 4NaOH \rightarrow 5MnO_2 + 2H_2O + 2KCl + 4NaCl, \quad (2.1)$$

However, it was indicated decades ago (Davies-Colley et al. 1984) that conditions of that kind lead to the formation of mineral called birnessite that contains manganese in two forms: (III) and (IV). Birnessite has a layered structure consisting of edge-shared MnO_6 octahedra with a wide interlayer space of about 7 Å (Liu et al. 2018). Other works indicate formation of manganese oxides mixture, too (Stone and Morgan 1984).

In attempts to obtain manganese dioxide from reaction (2.1), it was found that the resulting substance was not manganese dioxide. In various precipitation conditions, the sediment was formed with identical X-ray patterns. The example is shown in Figure 2.17, where the sample was obtained at temperature 80°C. X-ray structural analysis revealed in the obtained powders only weak reflections that can be attributed to manganese oxyhydroxide MnO(OH) and manganese(IV) dioxide MnO_2.

IR spectra of the obtained powders are shown in Figure 2.18. The absorption bands at 443 and 505 cm^{-1} can be attributed to Mn–O bond vibrations, and their presence was noted in MnO(OH) spectra (Crisostomo et al., 2007). Chemical analysis results are presented in Table 2.2. Data indicate mainly the presence of Mn(III) (36–41 wt%), as well as no more than 2 wt% of Mn(II) and no more than 6 wt% of Mn(IV). SEM images of the particles are shown in Figure 2.19.

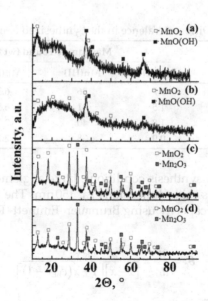

FIGURE 2.17
Diffraction diagrams of the synthesized powders: (a) after precipitation at $t = 80°C$; (b) after calcining for 1 hour at $t = 300°C$; (c) after calcining for 1 hour at $t = 400°C$; and (d) after calcining for 1 hour at $t = 700°C$.

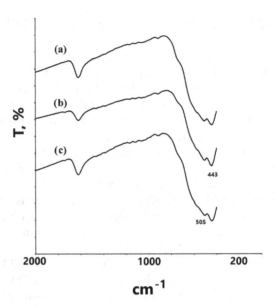

FIGURE 2.18
IR spectra of powders obtained from different concentrations of manganese solutions and at different temperatures: (a) 1 M and $t = 20°C$; (b) 1 M and $t = 80°C$; and (c) 0.1 M and $t = 80°C$. IR, infrared.

TABLE 2.2

Manganese Forms of Different Valence in the Synthesized Samples

Synthesis Conditions		Manganese Found (wt%)			
V_{all} (mL)	t (°C)	Mn(II)	Mn(III)	Mn(IV)	Overall Mn
750	20	1.7	36	6.3	44
750	80	1.5	41	6.5	49
75	20	1.8	40	2.2	44
75	80	0.9	38	2.1	41

Irrespective of the synthesis conditions, large agglomerates were formed, made up of particles with dimensions ca. 150 nm. The SSA was calculated from the monolayer capacity using Brunauer–Emmett–Teller (BET) equation (Naderi, 2015):

$$\theta_c = \frac{n}{n_m} = \frac{c_{BET} \cdot x}{[1-x][1+x(c_{BET}-1)]} \quad (2.2)$$

with relative pressure x:

$$x = \frac{P}{P_o} \quad (2.3)$$

FIGURE 2.19
SEM images of manganese(III) particles obtained at different concentrations of manganese solutions and different temperatures: (a) 1 M and 20°C; (b) 1 M and 60°C; (c) 1 M and 80°C; (d) 0.1 M and 20°C; (e) 0.1 M and 60°C; and (f) 0.1 M and 80°C. SEM, scanning electron microscopy.

For the particles obtained from 0.1 M solutions at temperatures 20°C and 80°C, SSAs were 360 and 210 m^2/g, respectively. However, SSA of the particles obtained from 1 M solutions at temperature 20°C was 160 m^2/g. This demonstrated that the SSA was larger when the manganese ions concentration in precursors was lower, but increase of the synthesis temperature caused decrease of SSA.

The samples of obtained powders underwent subsequent high-temperature calcination. X-ray diffraction measurement of the powders calcined at temperatures up to 300°C revealed practically no changes, as it was presented in Figure 2.17b. In contrast, diagrams of the powders calcined at 400°C intense reflections are seen in Figure 2.17c and d, which can be attributed to oxides Mn_2O_3 and MnO_2. Results of chemical analysis of calcined powders are collected in Table 2.3.

TABLE 2.3

Percentage of Different Oxidation States in Calcined Samples

	Mn Content (%)		
T_{calc} (°C)	Mn(II)	Mn(III)	Overall Mn (Mn(IV))
25	0.9	37.5	41 (2.6)
100	1.2	37.5	48 (9.3)
200	2.2	37.5	51.5 (11.8)
300	2.5	26	55 (26.5)
400	2.2	24	60 (33.8)
500	6.4	20.5	61.5 (34.6)
700	6.8	25	63.5 (31.7)

From the data, it can be derived that the samples calcined at lower temperatures up to 200°C contained similar proportion of Mn(III), but overall manganese content in a sample was increased. It is noteworthy that increase of overall manganese content took place in all temperature range from 25°C up to 700°C. Further heating above 300°C caused substantial decrease of Mn(III) content down to 20 wt% at 500°C, with simultaneous increase of Mn(II) content up to 6.4 wt%. However, the sample calcined at 700°C contained higher proportion of Mn(III). Figure 2.20 presents IR spectra diagrams of the calcined powders.

It can be noted from Figure 2.20a–c that the powders calcined at temperatures 100°C–300°C are basically identical. Weak step-resembling absorption bands are seen at 752 and 585 cm^{-1}, as well as absorption bands at 516 and 453 cm^{-1}. The latter is typical for the bond Mn–O in MnOOH molecule. On the other hand, the absorption band at 516 cm^{-1} can be attributed to Mn–O bond vibration in Mn_2O_3 (Kar et al. 2015), whereas the one at 585 cm^{-1} can be attributed to Mn–O bond vibration either in Mn_2O_3 or MnO_2 molecules (Khan et al. 2011). It is most probable that the absorption band at 752 cm^{-1} is the result of Mn–O bond vibration, too.

When the powders were calcined at higher temperatures of 400°C–500°C, their IR spectra revealed absorption bands at 706 to 703, 585, 516, and 463 cm^{-1} as it is shown in Figure 2.20d and e. The first ones, corresponding with 706 to 703 cm^{-1}, can be attributed to Mn–O bond vibration in MnO_2 (Khan et al. 2011),

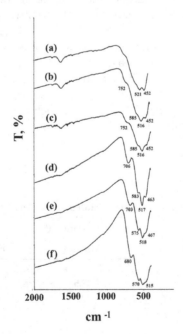

FIGURE 2.20
IR spectra of calcined samples for 1 hour at different temperatures: (a) 100°C; (b) 200°C; (c) 300°C; (d) 400°C; (e) 500°C; and (f) 700°C. IR, infrared.

whereas the last one at 463 cm⁻¹ to Mn–O bond vibration in MnOOH. It is noteworthy that for the sample calcined at 500°C, the absorption band intensity at 463 cm⁻¹ decreased, and it is present in the diagram as a weak step-resembling pulse. Furthermore, for the sample calcined at 700°C, the absorption band at 463 cm⁻¹ disappeared, but the bands at 680, 570, and 515 cm⁻¹ are present. The one corresponding with 680 cm⁻¹ can be attributed to Mn–O bond vibration in MnO_2 (Li et al. 2010), while the one at 570 cm⁻¹ can be attributed to Mn–O bond vibration either in Mn_2O_3 or MnO_2 molecules.

Thermogravimetric analysis results are presented in Figure 2.21. In the temperature range between 60°C and 350°C, main mass loss of 24 wt% takes place with its peak at 130°C. This may be attributed both to the removal of the adsorbed water from the particles surface and to the partial decomposition of manganese oxyhydroxide represented by the following reactions (Sampanthar et al. 2007):

$$4MnOOH + O_2 \rightarrow 4MnO_2 + 2H_2O \tag{2.4}$$

$$2MnO(OH) \rightarrow Mn_2O_3 + H_2O \tag{2.5}$$

Data obtained from chemical analysis and collected in Table 2.3 indicated that calcination at 100°C–200°C caused, on one hand, increased overall

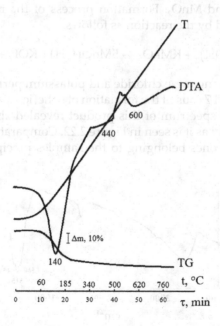

FIGURE 2.21
Thermogravimetric analysis of the dried sample of the powder obtained from manganese solution at $t = 80°C$, $V_{all} = 75$ mL. Curves: T – temperature, TG – mass changes, DTA – differential thermal analysis curve.

content of manganese after removal of the adsorbed water and, on the other hand, increased content of Mn(IV) as a result of reaction (2.2). Consequently, in the IR spectra of the samples calcined at 200°C, there were absorption bands corresponding with 752 and 585 cm^{-1}. Appearance of the absorption band at 516 cm^{-1} indicates that reaction (2.3) took place.

Subsequent heating caused further loss of mass indicated in the thermogravimetric diagram by the curve marked TG in Figure 2.21. The differential thermal analysis curve DTA exhibits thermal peak in the range of 350–550 with its maximum at 440°C. It corresponds with mass loss of 4% due to the manganese oxyhydroxide decomposition described by equation (2.2).

Chemical analysis presented in Table 2.3 shows the increase of Mn(IV) amount, whereas IR spectra reveal the absorption bands at 703 and 575 cm^{-1}. The absorption band at 453 cm^{-1} that corresponds with bond Mn–O vibrations in MnO(OH) becomes weaker. Thermal peak in the range of 500–800 with its maximum at 600°C (mass loss of 2%) corresponds with decomposition of manganese dioxide according to the following equation:

$$6MnO_2 \rightarrow 3Mn_2O_3 + 1.5O_2 \tag{2.6}$$

Thus, there is increase of Mn(III) in the examined samples.

Thus, the main product of the interaction between Mn^{2+} and MnO_4^{2-} in alkaline environment is MnO(OH). Its decomposition produces the mixture of oxides Mn_2O_3 and MnO_2. Formation process of the manganese oxyhydroxide is described by the reaction as follows:

$$4Mn(OH)_2 + KMnO_4 \rightarrow 5MnO(OH) + KOH + H_2O \tag{2.7}$$

Interaction of the manganese chloride and potassium permanganate in neutral environment pH 7 caused the formation of spherical particles with diameters 0.4–0.7 μm. IR spectrum of this product revealed absorption bands at 560, 513, and 458 cm^{-1} as it is seen in Figure 2.22. Comparative analysis of this spectrum with the ones belonging to the samples precipitated in alkaline

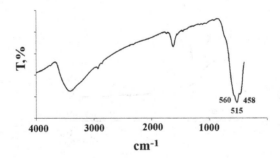

FIGURE 2.22
IR spectrum of the powder obtained from manganese chloride and potassium permanganate reaction in neutral environment pH 7. IR, infrared.

environment indicated that the shift of absorption band from 443 to 458 cm^{-1} accompanied by its intensity decreases. The absorption band at 560 cm^{-1} is attributed to Mn–O bond vibration either in Mn_2O_3 or in MnO_2 molecules. It can be assumed that neutral environment pH 7 results with decreased formation of manganese oxyhydroxide and thus increased proportion of obtained oxides.

2.2.2 Synthesis of Manganese Dioxide

MnO_2 particles were obtained after thermal decomposition of manganese nitrate at 200°C during 2 hours. Presence of the manganese dioxide was confirmed by the IR structural analysis. In the IR spectrum of the obtained powder shown in Figure 2.23, absorption bands are seen at 680, 650, 600, and 522 cm^{-1}. They can be attributed to Mn–O bond vibration in MnO_2.

Figure 2.24a presents SEM images of the samples obtained from reaction (2.5), and Figure 2.24b shows the samples from thermal decomposition. The latter consists of spherical nanoparticles below 50 nm. However, their SSA was very small, just 5.3 m^2/g.

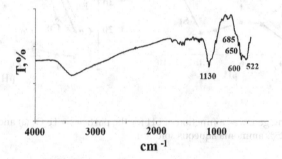

FIGURE 2.23
IR spectrum of the powder obtained from thermal decomposition of manganese nitrate at 200°C during 2 hours. IR, infrared.

FIGURE 2.24
SEM images of the particles obtained from ions Mn^{2+} and MnO_4^{2-} interaction at pH 7 (a) and from thermal decomposition of manganese nitrate at 200°C (b). SEM, scanning electron microscopy.

2.3 Sorption Characteristics of Iron Oxides and Manganese Dioxide and Oxyhydroxide

2.3.1 Iron Oxides

In Figures 2.25 and 2.26, diagrams of metal ions extraction percentage $E\%$ are shown for α-Fe_2O_3, γ-Fe_2O_3, and Fe_3O_4 particles as a function of pH. Figure 2.25 represents experiments where ammonia aqueous solution of 25% was used for pH regulation, whereas Figure 2.26 illustrates the case where 0.1 M solution of NaOH was used. In the range of pH between 5 and 9, the

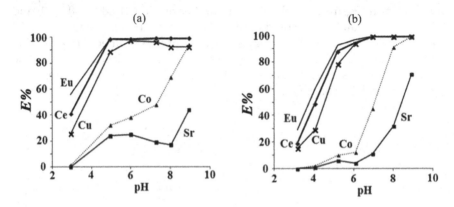

FIGURE 2.25
Extraction percentage $E\%$ as a function of pH for the particles α-Fe_2O_3 (a) and Fe_3O_4 (b) at pH regulation with 25% ammonia aqueous solution.

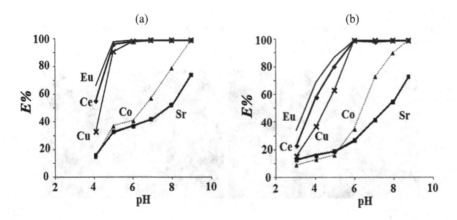

FIGURE 2.26
Extraction percentage as a function of pH for the particles α-Fe_2O_3 (a) and Fe_3O_4 (b) at pH regulation with 0.1 M solution of NaOH.

particles performed high ability to extract europium, cerium, and copper removing more than 95% of metal ions. Maximal values of $E\%$ were reached at pH 5.5 for both α-Fe_2O_3 and γ-Fe_2O_3, whereas in case of Fe_3O_4, maximal extraction percentage was reached at pH 6.5.

Moreover, iron oxide particles performed good selectivity toward copper, europium, and cerium, which is important in case of selective removal applications. Cobalt and strontium extraction percentage increased for higher pH values and became maximal at pH 9. Strontium exhibited small peak at pH 5.5 seen in Figure 2.25b.

According to the pH_{50} parameter, ions α-Fe_2O_3 can be ordered from smaller to larger pH: Eu^{3+} (pH_{50} 3) > Ce^{3+} (pH_{50} 3.2) > Cu^{2+} (pH_{50} 3.5) > Zn^{2+} (pH_{50} 5) > Cd^{2+} (pH_{50} 6) > Co^{2+} (pH_{50} 7.5) > Mn^{2+} (pH_{50} 8). In case of strontium ions, pH_{50} is reached at pH over 9. When the metals are ordered this way, the results stay in conformity with data reported by Takematsu (1979–1980). However, our values of pH_{50} parameter appear to be different. For instance, pH_{50} toward cerium is 5.8 with extraction percentage 100% reached at pH 7, whereas for zinc, it is 7.5 (Musić and Ristic 1988), and for copper, it is 5.5 (Benjamin and Leckie 1981). Shift of the pH_{50} parameter toward acidity can be attributed to the state of the sorbent surface dependent on its synthesis prehistory.

As it is seen from Figure 2.26, replacement of ammonia solution with NaOH solution had no substantial effect on the subsequent pH_{50} parameter. However, additional peak for strontium removal disappeared, and $E\%$ curve rises plainly up to 70% at pH 9.

It must be taken into account, however, that pH_{50} parameter is mainly a function of the solid-to-liquid ratio, with smaller effect of other variables. Kosmulski (2001, p. 355) indicated that according to systematic studies, pH_{50} parameter of metal cations decreases by one pH unit when the solid-to-liquid ratio increases by the order of magnitude, provided that the other variables remain unchanged.

Phase changes in Fe_2O_3 do not have noticeable effect on the extraction percentage from aqueous solutions. Thus, in case of γ-Fe_2O_3 phase, metal ions extraction dynamics is similar, as it is seen in Figure 2.27. Increase of pH caused increase of the extraction percentage up to maximal values of 90% at pH 5 for copper, cerium, and europium. Also in case of cobalt and strontium, extraction percentage increased for higher pH values and reached maximal 90% and 60%, respectively, at pH 9.

Figure 2.28 presents leaching of iron ions from the studies sorbent dependent on pH. It is known that Fe_3O_4 is better soluble in acid environment than Fe_2O_3. The measurement results demonstrated that iron concentration in the solutions after sorption for both examined oxides at pH 5 did not exceed maximal acceptable concentration (MAC_{Fe}^{3+}) of 0.3 mg/L (Mishukova et al. 2015). The World Health Organization indicated that in well water, iron concentrations below 0.3 mg/L were characterized as unnoticeable, whereas levels of 0.3–3 mg/L were found acceptable, especially for people drinking anaerobic well water (WHO 2003). Drinking Water Standards issued by the

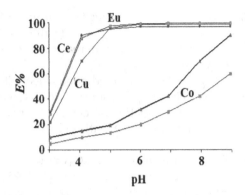

FIGURE 2.27
Extraction percentage E% as a function of pH for the particles γ-Fe_2O_3.

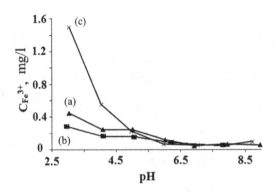

FIGURE 2.28
Iron concentration C_{Fe}^{3+} in the solution after 40 minutes: (a) α-Fe_2O_3, (b) γ-Fe_2O_3, and (c) Fe_3O_4.

Bureau of Indian Standards (IS: 10500: 1991) sets desirable limit at 0.3 mg/L and permissible limit at 1.0 mg/L. However, the European Union Drinking Water Directive (98/83/EC) prescribes limit value for iron as 200 µg/L (Environmental Protection Agency 2001).

To summarize, considering acceptable concentration of iron in drinking water as 0.3 mg/L, iron oxides may be successfully applied for its removal at pH above 5.

Figures 2.29–2.31 present adsorption isotherms of europium, cerium, copper, and cobalt. This sort of isotherms indicates that the sorbent surface is nonuniform. The results of calculation according to Langmuir model are shown in Table 2.4.

It should be noted that sorption capacity of Fe_2O_3 particles with SSA = 150 m^2/g is higher than that of Fe_3O_4-based sorbents with SSA = 130 m^2/g, which is true for all analyzed sorbed metals. The highest sorption capacity was observed for europium, and it was 21.3 and 19.7 mg/g on the Fe_2O_3 and

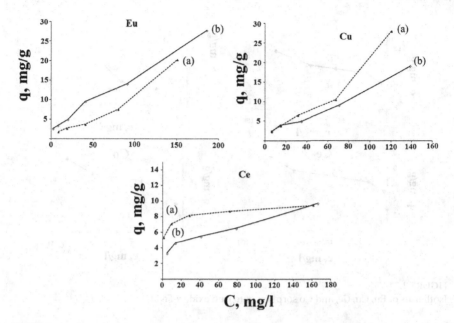

FIGURE 2.29
Isotherms of Eu, Cu, and Ce sorption onto iron oxides: (a) Fe_2O_3 and (b) Fe_3O_4.

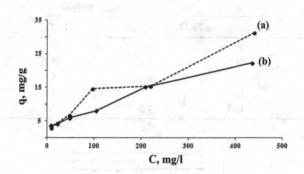

FIGURE 2.30
Cobalt sorption isotherms onto iron oxides: (a) Fe_2O_3 and (b) Fe_3O_4.

Fe_3O_4 particles, respectively. The smallest one took place in case of cerium and was 9.2 mg/g on Fe_2O_3 particles and 7.5 mg/g on Fe_3O_4 particles.

2.3.2 Manganese Oxyhydroxide and Dioxide

The experimental researches proved that the extraction percentage of MnO(OH) does not depend on the SSA, and it is over 90% in case of europium, cerium, cobalt, and strontium ions removal at pH 3–9. Figure 2.32 presents extraction percentage $E\%$ of metal ions onto MnO(OH) particles

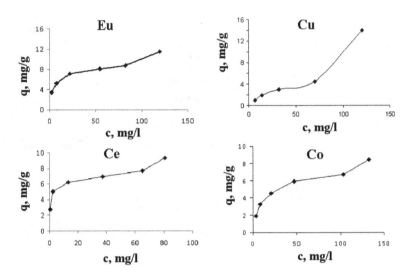

FIGURE 2.31
Isotherms of Eu, Cu, Ce, and Co sorption onto iron oxide γ-Fe$_2$O$_3$.

TABLE 2.4
Parameters of Sorption Isotherms onto Iron Oxides Fe$_2$O$_3$ and Fe$_3$O$_4$

	q_{max} (mg/g)		
	α-Fe$_2$O$_3$	Fe$_3$O$_4$	γ-Fe$_2$O$_3$
SSA (m^2/g)	150	130	35
Eu (pH 5)	21.3	19.7	8.8
Ce (pH 5)	9.2	8.9	7.1
Cu (pH 5)	15.7	14.6	6.5
Co (pH 6.5)	18.7	17.3	7.4

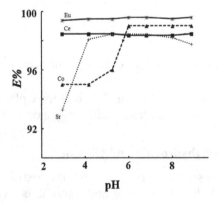

FIGURE 2.32
Extraction percentage E% of metal ions onto MnO(OH) particles at various pH.

with SSA = 160 m²/g. In case of europium and cerium, $E\%$ is over 98% at any pH. However, extraction of cobalt at pH below 6 decreases from 98% down to 95%, whereas $E\%$ of strontium decreases even down to 93%.

For the particles with different SSA, sorption capacity toward europium, cerium, strontium, and cobalt was determined. In Figure 2.33, there are sorption isotherms of these metals on the particles MnO(OH) with SSA = 360 m²/g.

In all cases, isotherms are quite similar, especially the ones representing strontium and cobalt. Sorption isotherm for cerium exhibits almost the same shape, but the values are of ca. 50% higher. Europium sorption isotherm at the beginning lays close to the lines 1 and 2 and then moves higher than the line 3. Similar curvature of isoterms can be noted for three samples of manganese oxyhydroxide with different SSAs.

Values of sorption capacity q_{max} of MnO(OH) particles with different SSA toward different metal ions are collected in Table 2.5. It is noteworthy that SSA has little effect on sorption capacity. For each sorbent, sorption capacity

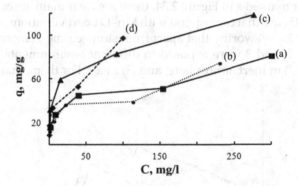

FIGURE 2.33
Sorption isotherms q_e of different metal ions at $T = 20°C$ on the MnO(OH) with SSA = 360 m²/g: (1) Sr; (2) Co; (3) Ce; and (4) Eu. SSA, specific surface area.

TABLE 2.5

Sorption Capacity of MnO(OH) Particles Toward Different Metals Calculated According to Langmuir Model

	SSA = 160 m²/g			SSA = 210 m²/g			SSA = 360 m²/g		
Metal	K_L	q_{max} (mg/g)	R^2 (%)	K_L	q_{max} (mg/g)	R^2 (%)	K_L	q_{max} (mg/g)	R^2 (%)
Eu(pH5)	-	-	-	0.02	49.5	96.64	22.78	49.1	96.67
Ce(pH5)	1.59	100.0	95.14	0.05	83.2	91.77	0.22	98.0	86.59
Sr(pH5)	0.73	45.2	88.61	-	-	-	0.30	50.6	91.47
Co(pH6.5)	-	-	-	0.03	20.0	96.24	1.19	32.0	90.93

SSA, specific surface area.

remains almost unchanged for different values of specific surface, and it can be noted that q_{max} is largest in case of cerium and smallest in case of cobalt.

During the experimental works, it was found that the storage process had negative effect on the MnO(OH) sorption capacity toward cobalt. The hypothesis was proposed that the decrease of sorption capacity can be attributed to the changes in substance composition. In order to verify this hypothesis, composition of the manganese oxyhydroxide was examined after storage in various conditions. Namely, the group of samples MnO(OH)-1 were stored for τ_{stor} = 1 year in the air environment, the samples MnO(OH)-2 were stored in argon atmosphere and MnO(OH)-3 were stored in high humidity. Some samples were taken from each group and checked after 3, 6, and 12 months. Before the measurement, the humid samples were dried in the air at room temperature for 24 hours.

X-ray diffraction (XRD) analysis of the samples indicated that, irrespective of storage conditions, all diagrams were similar to the one obtained for the freshly synthesized powder MnO(OH). Very weak reflections corresponding with MnO(OH) and MnO_2 are distinguishable. In the IR spectrum of the fresh powder marked a in Figure 2.34, there are clear main absorption bands at 443 and 505 cm^{-1} that correspond with Mn–O bond vibrations in MnO(OH) molecules. It is noteworthy that apart from storage time 12 months, samples MnO(OH)-1, 2, and 3 were exposed to different environmental conditions: (1) in water, (2) in inert atmosphere, and (3) in air. For them, main peaks can

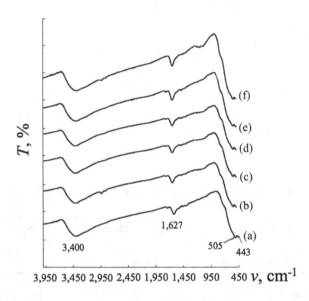

FIGURE 2.34
IR spectra of the samples: (a) freshly synthesized powder; (b) sample MnO(OH)-1 after 3 months; (c) sample MnO(OH)-1 after 6 months; (d) sample MnO(OH)-1 after 12 months; (e) sample MnO(OH)-2 after 12 months; and (f) sample MnO(OH)-3 after 12 months. IR, infrared.

be observed only at 443 and 505 cm^{-1}, attributed to Mn–O bond vibrations in MnO(OH) molecules.

Results of chemical analysis are collected in Table 2.6. They indicate that the main oxidation state of manganese is Mn(III). Variations of the Mn(II) and Mn(IV) content in the samples are omittable. However, the effect of the storage process on the decrease of SSA is substantial. The initial SSA for the samples MnO(OH)-1 and MnO(OH)-2 was 360 m^2/g, but after they were stored for 12 months, SSA decreased down to 266 and 258 m^2/g, respectively.

Thus, it was proved that storage of MnO(OH) powder for τ_{stor} = 12 months does not effect with its chemical composition changes.

Differentiation of the storage conditions had no effect on the extraction percentage, too. Figure 2.35 presents the $E\%$ results for strontium, europium, and cobalt removal onto freshly synthesized particles and onto the ones stored for 1 year in air. In the range pH 4–9, extraction percentage was above 90%, 98%, and 96%, respectively.

In order to determine maximal sorption capacity toward cobalt, strontium, and europium, sorption isotherms were drawn. Table 2.7 presents results calculated from the Langmuir model for cobalt ions. Substantial decrease of

TABLE 2.6

Chemical Analysis Results for Sample Group MnO(OH)-1

τ_{stor} (months)	Content of Mn (%)		
	Mn(II)	Mn(III)	Overall Mn (Mn(IV))
-	1.8	39.5	44.0 (2.7)
3	1.7	41.9	45.5 (1.9)
6	1.7	42.7	46.0 (1.6)
12	1.7	42.7	46.0 (1.6)

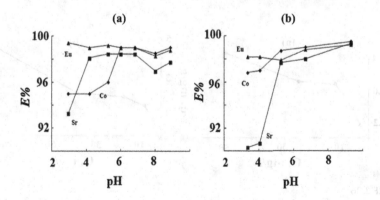

FIGURE 2.35
MnO(OH) extraction percentage $E\%$ versus pH: (a) freshly synthesized powder and (b) samples MnO(OH)-1 after being stored for 12 months.

TABLE 2.7
Sorption Characteristics of MnO(OH) Stored in Different Conditions

	Samples								
	MnO(OH)-1			MnO(OH)-2			MnO(OH)-3		
τ_{stor} (months)	K_L	q_{max} (mg/g)	R^2 (%)	K_L	q_{max} (mg/g)	R^2 (%)	K_L	q_{max} (mg/g)	R^2 (%)
Fresh	0.02	128.0	98.92	-	-	-	-	-	-
3	0.05	53.5	99.19	0.09	48.0	93.70	0.16	45.1	90.43
6	0.28	25.3	98.16	0.11	27.9	93.93	0.03	28.7	80.63
12	0.03	25.4	96.24	0.12	25.6	95.81	0.07	26.2	91.25

the sorption capacity toward cobalt was noted after 6 months, irrespective of the storage conditions. The capacity dropped from 128 down to 25 mg/g. However, further storage time had no effect on the sorption capacity of the MnO(OH) particles.

There is noticeable decrease of the sorption capacity toward europium and strontium after storage. Fresh powder exhibited capacity 108.7 mg/g, while after 1 year, its normal atmospheric conditions were decreased down to 48.8 mg/g toward europium and 27.9 mg/g toward strontium. Hence, it may be assumed that after first 6 months of storage, MnO(OH) particles exhibited substantial loss of their sorption capacity toward cobalt, europium, and strontium.

Figure 2.36 presents comparative isotherms for the cobalt sorption onto MnO_2 particles. They represent the particles obtained from decomposition of manganese nitrate at 200°C and the manganese oxyhydroxide calcined at 700°C. The former exhibited sorption capacity toward cobalt 2.5 mg/g, whereas the latter 4 mg/g.

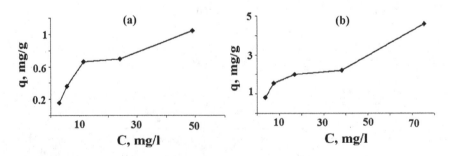

FIGURE 2.36
Sorption isotherms of cobalt at pH 5 onto (a) MnO_2 particles obtained from decomposition of manganese nitrate at 200°C and (b) the manganese oxyhydroxide calcined at 700°C.

2.4 Conclusions

To summarize the aforementioned experimental researches, it should be emphasized that formation process of iron oxide particles was examined in detail. In particular, effect of pH and precursor salt anions (chloride, nitrate, and hydrocarbonate anions) on the phase composition and morphological features of Fe_2O_3 was investigated in the process of ammonia precipitation from aqueous solutions.

Precipitation of iron(III) from nitrate solutions enables to obtain Fe_2O_3 particles made up of nanoparticles joined into larger agglomerates with dimensions over 50 µm, irrespective of the synthesis pH.

The powder containing separate Fe_2O_3 particles of cubic and oval shapes is formed in presence of chloride anions. However, carbonate anion does not affect the morphology of particles. Hence, the form of the particles synthesized through thermal decomposition of iron(III) salts is highly dependent on the nature of anions of the iron(III) salt precursors. It is possible to control anion composition and thus to obtain iron oxide particles with desired functional characteristics.

Furthermore, the effect of the precipitation conditions, such as temperature and iron concentration in a solution, on the phase content, as well as the particle dimensions and magnetization, was analyzed. Comparative analysis of the extraction percentage and sorption capacity of magnetite and hematite toward cobalt was performed. It was found that the magnetite phase formation is dependent substantially on the precipitation temperature. The higher the temperature, the higher the magnetite content in samples. When synthesis was performed at 90°C in 0.15 M concentration of iron, the powder with 100 wt% Fe_3O_4 magnetite was obtained. It exhibited magnetization ca. 70 A·m²/kg. Increase of initial concentration of iron in the solution resulted with slight decrease of magnetization.

The obtained particle shape was spherical irrespective of the precipitation conditions, and their dimensions were between 7 and 15 nm. Larger particles with higher magnetization were noted to be formed in higher precipitation temperatures and lower iron concentrations.

During the interaction of ions Mn^{2+} and MnO_4^{2-} in alkaline environment, the main component is manganese oxyhydroxide MnO(OH), not manganese dioxide. It was proved by the IR spectrometry. SSA of the particles increases, when the synthesis temperature is lower, and concentration of Mn^{2+} ions is smaller, and the obtained SSA is located in the range between 160 and 360 m²/g. Particles with the largest SSA of 360 m²/g can be formed in the 0.1M solution at t = 20°C.

Thermal decomposition of MnO(OH) proceeds in several stages. In the first stage, at the temperature between 60°C and 350°C, the adsorbed water is removed from the surfaces of particles. Partial oxidation of manganese oxyhydroxide to MnO_2 takes place. In the second stage, at the temperature between 350°C and 550°C, main oxidation of MnO(OH) to manganese dioxide is completed. Further heating above 600°C will cause decomposition of manganese dioxide and formation of Mn_2O_3.

Compared with the particles of iron oxides, manganese oxyhydroxide particles perform substantially better sorption capacity toward the metals. Sorption capacity of iron oxides does not depend on phase composition but is increased when SSA is larger. On the other hand, changes in particles of manganese oxohydroxide SSA have no effect on their sorption capacity.

However, sorption properties of the manganese oxyhydroxide particles worsen during the first 6 months of storage. Among others, its sorption capacity toward cobalt drops down to 20% of initial value after 6 months and then remains almost constant at the level of 25 mg/g. In case of sorption capacity toward strontium and europium, it decreases 50% during 12 months and then remains 30 and 50 mg/g, respectively.

Final Remarks

There are no doubts that sorption materials have wide perspectives of application in contemporary complex systems of water treatment. Modern trends of their development are dependent on the need of appropriate water supply, necessary for life of present and future generations. Thus, the main problem of the technological systems is to work out ecologically safe technologies with minimal amounts of wastes. The researches are focused on the improvement of existing materials and development of new ones of high sorption efficiency. These materials are to be used in complex and selective processes of wastewater treatment in the context of ecology, energy, and resource savings, as well as social life.

One of the solutions leading to achievement of proper materials is the development of methods of controlled synthesis of particles with presumed desirable characteristics. New technologies can be developed after thorough analysis of a wide range of factors that have effect on the particles structure and morphology formation, such as temperature, pH, composition, activation method, and so on. These technologies may be applied also to production of other materials, not only for sorption purposes.

Physical and chemical characteristics of the metal sulfides, iron oxides, and manganese oxyhydroxide allow their application as sorbents. They may be used for heavy metals and radionuclides uptake from wastewater practically down to any residual concentration at pH above 5. The controlled inorganic synthesis allows forming of sorbents particles with certain morphological characteristics, which effects with increase of sorption capacity even several times.

It appears to be the most promising to apply manganese oxyhydroxide as a sorbent. It is able to uptake heavy metal ions and radionuclides with high efficiency above 95% (Sofronov et al. 2019), including isotopes ^{90}Sr and ^{137}Cs (Krasnopyorova et al. 2017), and perform higher sorption capacity than that of metal sulfides and iron oxides.

The research results presented in this work are aimed on the controlled synthesis of inorganic particles with presumed structural, morphological, and functional properties.

Acknowledgments

The authors express their gratitude for cooperation to everybody who contributed to the researches, especially to chemistry doctors Baumer V.N. and Puzan A.N. for X-ray structural analysis, to Bunina Z.Yu., Gudzenko L.V., and chemistry doctor Shcherbakov I.B.-Kh. for chemical analysis, to Mateichenko P.V. for microstructural analysis, to chemistry doctor Beda A.A. for specific surface determination, and to physics and mathematics doctor Katrunov K.A. for optical researches.

This chapter is made Open Access through funding by Precision Machine Parts Poland Sp. z o.o.

References

Abdullah, N.H., Z. Zainal, S. Silong, M.I.M. Tahir, K.B. Tan, and S.K. Chang. 2016. Synthesis of zinc sulphide nanoparticles from thermal decomposition of zinc N-ethyl cyclohexyl dithiocarbamate complex. *Materials Chemistry and Physics* 173:33–41.

Acton, Q.A. (ed.), 2013. *Zinc Compounds—Advances in Research and Application: 2013 Edition*. Atlanta: Scholarly Editions.

Adler, H.H., and P.F. Kerr. 1965. Variations in infrared spectra, molecular symmetry and site symmetry of sulfate minerals. *American Mineralogist* 50(1–2):132–47.

Afridi, M.N., W.-H. Lee, and J.-O. Kim. 2019. Effect of phosphate concentration, anions, heavy metals, and organic matter on phosphate adsorption from wastewater using anodized iron oxide nanoflakes. *Environmental Research* 171:428–36.

Aguilar, O., F. Tzompantzi, R. Pérez-Hernández, R. Gómez, and A. Hernández-Gordillo. 2017. Novel preparation of ZnS from $Zn_5(CO_3)_2(OH)_6$ by the hydro- or solvothermal method for H_2 production. *Catalysis Today* 287:91–8.

Ahmed, K.A.M. 2016. Exploitation of $KMnO_4$ material as precursors for the fabrication of manganese oxide nanomaterials. *Journal of Taibah University for Science* 10(3):412–29.

Ai, Zh., Y. Cheng, L. Zhang, and J. Qiu 2008. Efficient Removal of Cr(VI) from Aqueous Solution with $Fe-Fe_2O_3$ Core-Shell Nanowires. *Environmental Science & Technology* 42(18):6955–60.

Alby, D., C. Charnay, M. Heran, B. Prelot, and J. Zajac. 2018. Recent developments in nanostructured inorganic materials for sorption of cesium and strontium: Synthesis and shaping, sorption capacity, mechanisms, and selectivity – A review. *Journal of Hazardous Materials* 344:511–530.

Allan, K.F., S. Labib, and M. Holeil. 2015. Synthesis and characterization of iron sulfide powders and its application for sorption of europium radionuclides. *Desalination and Water Treatment* 51:1–15.

Anand, K.V., M.K. Chinnu, R. Mohan Kumar, R. Mohan, and R. Jayavel. 2009. Formation of zinc sulfide nanoparticles in HMTA matrix. *Applied Surface Science* 255(21):8879–82.

Balsley, S.D., P.V. Brady, J.L. Krumhansl, et al. 1996. Iodide retention by metal sulfide surfaces: Cinnabar and chalcocite. *Environmental Science & Technology* 30(10):3025–7.

Baltas, H., M. Sirin, E. Gökbayrak, and A.E. Ozcelik. 2020. A case study on pollution and a human health risk assessment of heavy metals in agricultural soils around Sinop province, Turkey. *Chemosphere* 241:125015.

Baranov, D.A., and S.P. Gubin. 2009. Radioelectronics: Nanosystems. *Information Technologies* 1(1–2):129. (in Russian).

Barzyk, W., A. Kowal, and A. Pomianowski. 2002. Noble metal (Ag, Au) cementation on non-stoichiometric cuprous sulphide grains. *Colloids and Surfaces A: Physicochemical and Engineering Aspects* 208:321–35.

Baykal, A., Y. Köseolu, and M. Senel. 2007. Low temperature synthesis and characterization of Mn_3O_4 nanoparticles. *Central European Journal of Chemistry* 5(1):169–76.

Behboudnia, M., M.H. Majlesara, and B. Khanbabaee. 2005. Preparation of ZnS nanorods by ultrasonic waves. *Materials Science and Engineering B* 122:160–3.

Benjamin, M.M. and J.O. Leckie. 1981. Multiple-site adsorption of Cd, Cu, Zn, and Pb on amorphous iron oxyhydroxide. *Journal of Colloid and Interface Science* 79(1):209–21.

Bhateria, R., and R. Singh. 2019. A review on nanotechnological application of magnetic iron oxides for heavy metal removal. *Journal of Water Process Engineering* 31:100845.

Biswas, S., T. Ghoshal, S. Kar, S. Chakrabarti, and S. Chaudhuri. 2008. ZnS nanowire arrays: Synthesis, optical and field emission properties. *Crystal Growth & Design* 8(7):2171–6.

Bol, A.A., J. Ferwerda, J.A. Bergwerff, and A. Meijerink. 2002. Luminescence of nanocrystalline $ZnS:Cu^{2+}$. *Journal of Luminescence* 99:325–34.

Bostick, B.C., and S. Fendorf. 2003. Arsenite sorption on troilite (FeS) and pyrite (FeS_2). *Geochimica et Cosmochimica Acta* 67(5):909–21.

Bower, K.E., Y.A. Barbanel, Y.G. Shreter, and G.W. Bohnert (eds.). 2002. *Polymers, Phosphors, and Voltaics for Radioisotope Microbatteries*. Boca Raton, FL: CRC Press.

Brittain, H.G. (ed.) 1993. *Analytical Profiles of Drug Substances and Excipients*, Vol. 22. San Diego, CA: Academic Press Inc.

Bulgakova, A.V., D.S. Sofronov, P.V. Mateichenko, V.N. Baumer, A.A. Beda, and V.A. Chebanov. 2016. The effect of the precipitation conditions on the morphology and the sorption properties of CuS particles. *Protection of Metals and Physical Chemistry of Surfaces* 52(3):448–53.

Bulgakova, A.V., D.S. Sofronov, P.V. Mateychenko, V.N. Baumer, A.A. Beda, and V.A. Chebanov. 2016. Effect of the precipitation conditions on morphology and sorption characteristics of CuS. *Surface Physics and Chemistry and Materials Protection* 52(3):295–300. (in Russian).

Byrn, S.R., G. Zografi, and X. Chen. 2017. *Solid-State Properties of Pharmaceutical Materials*. Hoboken, NJ: John Wiley & Sons, Inc.

Cao, Ch.Y., J. Qu, W.-Sh. Yan, J.-F. Zhu, Z.-Y. Wu, and W.-G. Song. 2012. Low-cost synthesis of flowerlike α-Fe_2O_3 nanostructures for heavy metal ion removal: Adsorption property and mechanism. *Langmuir* 28(9):4573–9.

Cao, J., Y. Zhu, K. Bao, L. Shi, Sh. Liu, and Y. Qian. 2009. Microscale Mn_2O_3 hollow structures: Sphere, cube, ellipsoid, dumbbell, and their phenol adsorption properties. *Journal of Physical Chemistry C* 113(41):17755–60.

Cao, Sh.W., and Y.J. Zhu. 2008. Hierarchically nanostructured α-Fe_2O_3 hollow spheres: Preparation, growth mechanism, photocatalytic property, and application in water treatment. *Journal of Physical Chemistry C* 112(16):6253–7.

Chanu, T.I., Dh. Samanta, A. Tiwari, and S. Chatterjee. 2017. Effect of reaction parameters on photoluminescence and photocatalytic activity of zinc sulfide nanosphere synthesized by hydrothermal route. *Applied Surface Science* 391(Part B):548–56.

Cheera, P., S. Karlapudi, G. Sellola, and V. Ponneri. 2016. A facile green synthesis of spherical Fe_3O_4 magnetic nanoparticles and their effect on degradation of methylene blue in aqueous solution. *Journal of Molecular Liquids* 221:993–8.

Chen, D., K. Tang, G. Shen, et al. 2003. Microwave-assisted synthesis of metal sulfides in ethylene glycol. *Materials Chemistry and Physics* 82:206–9.

References

Chen, J., J. Wang, G. Zhang, Q. Wu, and D. Wang. 2018. Facile fabrication of nanostructured cerium-manganese binary oxide for enhanced arsenite removal from water. *Chemical Engineering Journal* 334:1518–26.

Cheng, Ch., G. Xu, H. Zhang, J. Cao, P. Jiao, and X. Wang. 2006. Low-temperature synthesis and optical properties of wurtzite ZnS nanowires. *Materials Letters* 60:3561–4.

Cheraneva, L.G., M.A. Shvetsova, and V.V. Volkhin. 2009. Impact of the sorbents synthesis conditions on the structure and characteristics of metals sulfides. *Bulletin of Perm National Research Polytechnic Institute: Chemical and Bio Technology* 9(1):19–25. (in Russian).

Coles, C.A., S.R. Rao, and R.N. Yong. 2000. Lead and cadmium interactions with mackinawite: Retention mechanisms and the role of pH. *Environmental Science & Technology* 34(6):996–1000.

Con, T.H., Ph. Thao, T.X. Dai, and D.K. Loan. 2013. Application of nano dimensional MnO_2 for high effective sorption of arsenic and fluoride in drinking water. *Environmental Sciences* 1(1–2):69–77.

Corrado, C., Y. Jiang, F. Oba, M. Kozina, F. Bridges, and J.Z. Zhang. 2009. Synthesis, structural, and optical properties of stable ZnS:Cu, Cl nanocrystals. *Journal of Physical Chemistry A* 113(16):3830–9.

Crisostomo, V.-M.B., J.K. Ngala, S. Alia, et al. 2007. New synthetic route, characterization, and electrocatalytic activity of nanosized manganite. *Chemistry of Materials* 19(7):1832–39.

Davies-Colley, R.J., P.O. Nelson, and K.J. Williamson. 1984. Copper and cadmium uptake by estuarine sedimentary phases. *Environmental Science & Technology* 18(7):491–9.

Dąbrowski, A., and V.A. Tertykh (eds.). 1996. *Adsorption on New and Modified Inorganic Sorbents*. Amsterdam: Elsevier.

Deng X., Zh. Huang, W. Wang, and R.N. Davé. 2016. Investigation of nanoparticle agglomerates properties using Monte Carlo simulations. *Advanced Powder Technology* 27(5):1971–9.

Dhara, A., S. Sain, S. Das, and S.K. Pradhan. 2018. Microstructure, optical and electrical characterizations of Mn doped ZnS nanocrystals synthesized by mechanical alloying. *Materials Research Bulletin* 97:169–75.

Dong L., Zh. Zhu, H. Ma, Y. Qiu, and J. Zhao. 2010. Simultaneous adsorption of lead and cadmium on MnO_2-loaded resin. *Journal of Environmental Sciences* 22(2):225–9.

Dong, L., Y. Chu, and Y. Zhang. 2007. Microemulsion-mediated solvothermal synthesis of ZnS nanowires. *Materials Letters* 61:4651–4.

Dragan, E.S. (ed.) 2014. *Advanced Separations by Specialized Sorbents*. Boca Raton, FL: CRC Press.

Dubey, Sh., S. Banerjee, S.N. Upadhyay, and Y.Ch. Sharma. 2017. Application of common nano-materials for removal of selected metallic species from water and wastewaters: A critical review. *Journal of Molecular Liquids* 240:656–77.

Edrah, S. 2010. Synthesis, characterization and biological activities of ureas and thioureas derivatives. *Journal of Applied Sciences Research* 4(8):1014–8.

Egorov, N.B., L.P. Eremin, A.M. Larionov, and V.F. Usov. 2010. Transformations of the thiosulfate-thiourea lead complexes under heating. *Bulletin of Tomsk Polytechnic University* 317(3):99–102. (in Russian).

Environmental Protection Agency. 2001. *Parameters of Water Quality: Interpretation and Standards*. Wexford: Environmental Protection Agency.

Fang, X., T. Zhai, U.K. Gautam, et al. 2011. ZnS nanostructures: From synthesis to applications. *Progress in Materials Science* 56:175–287.

Feng, X.H., L.M. Zhai, W.F. Tan, F. Liu, and J.Z. He. 2007. Adsorption and redox reactions of heavy metals on synthesized Mn oxide minerals. *Environmental Pollution* 147:366–73.

Fu, F., and Q. Wang. 2011. Removal of heavy metal ions from wastewaters: A review. *Journal of Environmental Management* 92(3):407–18.

Furniss, B.S., A.J. Hannaford, P.W.G. Smith, and A.R. Tatchell (eds.). 1989. *Vogel's Textbook of Practical Organic Chemistry*, 5th edn. Harlow: Longman Scientific & Technical.

Ge, S., X. Shi, K. Sun, et al. 2009. Facile hydrothermal synthesis of iron oxide nanoparticles with tunable magnetic properties. *Journal of Physical Chemistry C* 113(31):13593–9.

Geckeler, K., and H. Nishide (eds.). 2010. *Advanced Nanomaterials*. Weinheim: Wiley-VCH.

Geng, B., J. Ma, and F. Zhan. 2009. A solution phase thermal decomposition molecule precursors route to ZnS:Cu2+ nanorods and their optical properties. *Materials Chemistry and Physics* 113:534–8.

Gevorkyan, E.S., M. Rucki, V.A. Chishkala, M.V. Kislitsa, Z. Siemiatkowski, and D. Morozow. 2019. Hot pressing of tungsten monocarbide nanopowder mixtures by electroconsolidation method. *Journal of Machine Construction and Maintenance* 113(2):67–73.

Gong, Y., J. Tang, and D. Zhao. 2016. Application of iron sulfide particles for groundwater and soil remediation: A review. *Water Research* 89:309–20.

Gupta, K.M., and N. Gupta. 2016. *Advanced Semiconducting Materials and Devices*. Cham: Springer.

Guttikunda, S.K., K.A. Nishadh, and P. Jawahar. 2019. Air pollution knowledge assessments (APnA) for 20 Indian cities. *Urban Climate* 27:124–41.

Hamilton-Taylor, J., W. Davison, and K. Morfett. 1996. The biogeochemical cycling of Zn, Cu, Fe, Mn, and dissolved organic C in a seasonally anoxic lake. *Limnology and Oceanography* 41(3):408–18.

Han, R., W. Zou, Y. Wang, and L. Zhu. 2007. Removal of uranium(VI) from aqueous solutions by manganese oxide coated zeolite: Discussion of adsorption isotherms and pH effect. *Journal of Environmental Radioactivity* 93(3):127–43.

Herrmann J., W. Knoche, and R. Neugebauer. 1995. Hydrolysis of thiamine. *Journal of the Chemical Society. Perkin Transactions Part 2* 3:463–8.

Hong, R.Y., T.T. Pan, and H.Z. Li. 2006. Microwave synthesis of magnetic Fe_3O_4 nanoparticles used as a precursor of nanocomposites and ferrofluids. *Journal of Magnetism and Magnetic Materials* 303:60–8.

Hurma, T. 2016. Effect of cerium incorporation on the structural and optical properties of CdS film. *Optik – International Journal for Light and Electron Optics* 127(22):10670–5.

Jaffres, A., D. Bregiroux, D. Reekie, and R. Shears. 2017. Morphological control of ZnS nanopowders by different capping molecules. *Materials Letters* 209:539–542.

Jaspal, D., and A. Malviya. 2020. Composites for wastewater purification: A review. *Chemosphere* 246:125788.

References

Jayalakshmi, M., and M.M. Rao. 2006. Synthesis of zinc sulphide nanoparticles by thiourea hydrolysis and their characterization for electrochemical capacitor applications. *Journal of Power Sources* 157(1):624–9.

Jeong, H.Y., B. Klaue, J.D. Blum, et al. 2007. Sorption of mercuric ion by synthetic nanocrystalline mackinawite (FeS). *Environmental Science & Technology* 41(22):7699–705.

Jian, W., J. Zhuang, D. Zhang, J. Dai, W. Yang, and Y. Bai. 2006. Synthesis of highly luminescent and photostable ZnS:Ag nanocrystals under microwave irradiation. *Materials Chemistry and Physics* 99:494–7.

Jiang, Ch., W. Zhang, G. Zou, W. Yu, and Y. Qian. 2007. Hydrothermal synthesis and characterization of ZnS microspheres and hollow nanospheres. *Materials Chemistry and Physics* 103, 24–7.

Jothibas, M., S. Johnson Jeyakumar, C. Manoharan, I.K. Punithavathy, P. Praveen, and J.P. Richard. 2017. Structural and optical properties of zinc sulphide nanoparticles synthesized via solid state reaction method. *Journal of Materials Science: Materials in Electronics* 28:1889–94.

Jothibas, M., C. Manoharan, S. Johnson Jeyakumar, P. Praveen, I.K. Punithavathy, and J.P. Richard. 2018. Synthesis and enhanced photocatalytic property of Ni doped ZnS nanoparticles. *Solar Energy* 159:434–43.

Kar, P., S. Sardar, S. Ghosh, et al. 2015. Nano surface engineering of Mn_2O_3 for potential light-harvesting application. *Journal of Materials Chemistry C* 3:8200–11.

Katrunov, K., D. Sofronov, N. Starzhynskiy, et al. 2010. Quantum-size effects in nanosize particles ZnS and CdS obtained by precipitation from alkaline solutions. *Technology, Instrumentation and Production of Electonic Technique* 28(2):3–7. (in Russian).

Khan, Y., Sh.Kh. Durrani, M. Mehmood, and M.R. Khan. 2011. Mild hydrothermal synthesis of γ-MnO_2 nanostructures and their phase transformation to α-MnO_2 nanowires. *Journal of Materials Research* 26(17):2268–75.

Kim, D., K.-D. Min, J. Lee, J.H. Park, and J.H. Chun. 2006. Influences of surface capping on particle size and optical characteristics of ZnS:Cu nanocrystals. *Materials Science and Engineering B* 131:13–7.

Kim, J.H., J.G. Kim, J. Song, et al. 2018. Investigation of the growth and in situ heating transmission electron microscopy analysis of Ag_2S-catalyzed ZnS nanowires. *Applied Surface Science* 436:556–61.

Kitayev, G.A., V.F. Markov, and L.N. Maskayeva. 1989. Investigations on chalcogenide films of solid solutions CdXPb1-XS. *Bullettin of SU Academy of Sciences: Inorganic Materials* 26(8):1262–4 (in Russian).

Knunyantz, I.L. 1983. Chemical Ecyclopaedic Dictionary. Moscow: Soviet Encyclopedia Publisher (in Russian).

Kolida, Yu.Ya., A.S. Antonova, T.N. Kropacheva, and V.I. Kornev. 2014. Magnetic iron oxides as sorbents of hard metals cations. *Bulletin of Udmurt University* 4:52–61 (in Russian).

Kosmulski, M. 2001. *Chemical Properties of Material Surfaces*. New York; Basel: Marcel Dekker AG.

Krasnopyorova, A.P., N.V. Efimova, G.D. Yuhno, and D.S. Sofronov. 2017. Impact of ion force and solution acidity on sorption capability of MnO(OH) toward radionuclides ^{90}Sr, ^{137}Cs and ^{60}Co. *Vestnik of Novgorod State University* 103(3):22–5. (in Russian)

Labiadh, H., B. Sellami, A. Khazri, W. Saidani, and S. Khemais. 2017. Optical properties and toxicity of undoped and Mn-doped ZnS semiconductor nanoparticles synthesized through the aqueous route. *Optical Materials* 64:179–86.

Lavrynenko, S., A.G. Mamalis, D. Sofronov, A. Odnovolova, and V. Starikov. 2018. Synthesis features of iron oxide nanopowders with high magnetic and sorption properties. *Materials Science Forum* 915:116–20.

Lee, S., D. Song, D. Kim, et al. 2004. Effects of synthesis temperature on particle size/shape and photoluminescence characteristics of ZnS:Cu nanocrystals. *Materials Letters* 58:342–6.

Li, F., J. Wu, Q. Qin, Zh. Li, and X. Huang. 2010. Facile synthesis of γ-MnOOH micro/nanorods and their conversion to β-MnO_2, Mn_3O_4. *Journal of Alloys and Compounds* 492(1–2):339–46.

Li, W., D. Li, Zh. Chen, et al. 2008. High-efficient degradation of dyes by ZnxCd1-xS solid solutions under visible light irradiation. *Journal of Physical Chemistry C* 112:14943–7.

Li, Y., X. He, and M. Cao. 2008. Micro-emulsion-assisted synthesis of ZnS nanospheres and their photocatalytic activity. *Materials Research Bulletin* 43:3100–10.

Liu, C., Y. Chen, W. Huang, Y. Situ, and H. Huang. 2018. Birnessite manganese oxide nanosheets assembled on Ni foam as high-performance pseudocapacitor electrodes: Electrochemical oxidation driven porous honeycomb architecture formation. *Applied Surface Science* 458:10–7.

Liu, J., K.T. Valsaraj, I. Devai, and R.D. DeLaune, 2008. Immobilization of aqueous Hg(II) by mackinawite (FeS). *Journal of Hazardous Materials* 157:432–40.

Liu, T., L. Xue, and X. Guo, 2015. Study of Hg^0 removal characteristics on Fe_2O_3 with H_2S. *Fuel* 160:189–95.

Loganathan, P., and R.G. Burau. 1973. Sorption of heavy metal ions by a hydrous manganese oxide. *Geochimica et Cosmochimica Acta* 37(5), 1277–93.

Lompe, K.M., D. Menard, and B. Barbeau. 2017. The influence of iron oxide nanoparticles upon the adsorption of organic matter on magnetic powdered activated carbon. *Water Research* 123:30–9.

Mallakpour, Sh., and F. Motirasoul. 2019. Cross-linked poly(vinyl alcohol)/modified α-manganese dioxide composite as an innovative adsorbent for lead(II) ions. *Journal of Cleaner Production* 224:592–602.

Manoharan, S., S. Goyal, M.L. Rao, M.S. Nair, and A. Pradhan. 2001. Microwave synthesis and characterization of doped ZnS based phosphor materials. *Materials Research Bulletin* 36:1039–47.

Marcotrigiamo, G., G. Peyronel, and R. Battisturzi. 1972. Kinetics of desulphuration of thiourea in sodium hydroxide studied by chromatographic method. *Journal of the Chemical Society. Perkin Transactions Part 2* 11:1539–41.

Mateleshko, N., V. Mitsa, and S. Sikora. 2004. Electrooptical properties of DC electroluminescent ZnS:Mn, Cu powder panels with chalcogenide glass intermediate layer. *Journal of Optoelectronics and Advanced Materials* 6(1):329–32.

McCloy, J.S., and R.W. Tustison. 2013. *Chemical Vapor Deposited Zinc Sulfide*. Washington: SPIE Press.

Mendil, R., Z. Ben Ayadi, and K. Djessas. 2016. Effect of solvent medium on the structural, morphological and optical properties of ZnS nanoparticles synthesized by solvothermal route. *Journal of Alloys and Compounds* 678:87–92.

Mishukova, T.G., A.A. Osipov, and I.A. Salnikov. 2015. Determination of the microelements content in drinking waters of the Orenburg Region. *Bulletin of Orenburg State University* 185(10):303–7 (in Russian).

Moore, G.L. 1989. *Introduction to Inductively Coupled Plasma Atomic Emission Spectrometry.* Amsterdam: Elsevier.

Morse, J.W., and T. Arakaki. 1993. Adsorption and coprecipitation of divalent metals with mackinawite (FeS). *Geochimica et Cosmochimica Acta* 57(15):3635–40.

Mullet, M., S. Boursiquot, and J.J. Ehrhardt. 2004. Removal of hexavalent chromium from solutions by mackinawite, tetragonal FeS. *Colloids and Surfaces A Physicochemical and Engineering Aspects* 244(1):77–85.

Muraleedharan, K., V.K. Rajan, and V.M. Abdul Mujeeb. 2015. Green synthesis of pure and doped semiconductor nanoparticles of ZnS and CdS. *Transactions of Nonferrous Metals Society of China* 25(10):3265–70.

Murray, J.W. 1975. The interaction of cobalt with hydrous manganese dioxide. *Geochimica et Cosmochimica Acta* 39(5):635–47.

Murugadoss, G. 2013. Synthesis and photoluminescence properties of zinc sulfide nanoparticles doped with copper using effective surfactants. *Particuology* 11(5):566–73.

Musić, S. 1985. Sorption of chromium(III) and chromium(VI) on lead sulfide. *Journal of Radioanalytical and Nuclear Chemistry* 91(2):337–47.

Musić, S. and M. Ristić. 1988. Adsorption of trace elements or radionuclides on hydrous iron oxides. *Journal of Radioanalytical and Nuclear Chemistry* 120(2):289–304.

Naderi, M. 2015. Surface area: Brunauer–Emmett–Teller (BET). In: S. Tarleton (ed.) *Progress in Filtration and Separation*, pp. 585–608. London: Academic Press/Elsevier.

Nagpal, M., and R. Kakkar. 2019. Use of metal oxides for the adsorptive removal of toxic organic pollutants. *Separation and Purification Technology* 211:522–39.

Neto, J.O.M., C.R. Bellato, and D.C. Silva. 2019. Iron oxide/carbon nanotubes/chitosan magnetic composite film for chromium species removal. *Chemosphere* 218:391–401.

Niazi, N.K., and E.D. Burton. 2016. Arsenic sorption to nanoparticulate mackinawite (FeS): An examination of phosphate competition. *Environmental Pollution* 218:111–7.

Nishikida, K., E. Nishio, and R.W. Hannah. 1995. *Selected Applications of Modern FT-IR Techniques.* Tokyo: Kodansha.

Nouh, S.A., M. Amin, M. Gouda, and A. Abd-Elmagid. 2015. Extraction of uranium(VI) from sulfate leach liquor after iron removal using manganese oxide coated zeolite. *Journal of Environmental Chemical Engineering* 3:523–8.

Nyquist, R.A., and R.O. Kagel. 1971. *Infrared Spectra of Inorganic Compounds.* New York; London: Academic Press.

Ociński, D., I. Jacukowicz-Sobala, J. Raczyk, and E. Kociołek-Balawejder. 2014. Evaluation of hybrid polymer containing iron oxides as As(III) and As(V) sorbent for drinking water purification. *Reactive and Functional Polymers* 83:24–32.

Odnovolova, A.M., D.S. Sofronov, P.V. Mateichenko, et al. 2014. Role of anion composition of aqueous solution in forming morphology and surface of particles Fe_2O_3 in the course of deposition and their sorption properties. *Russian Journal of Applied Chemistry* 87(8):1060–4.

Odnovolova, A.M., D.S. Sofronov, A. Puzan, et al. 2015. Formation characteristics of Fe_3O_4 magnetic particles precipitated from aqueous solutions and their sorption properties. *Functional materials* 22(4):475–81.

Pandi K., S. Periyasamy, and N. Viswanathan. 2017. Remediation of fluoride from drinking water using magnetic iron oxide coated hydrotalcite/chitosan composite. *International Journal of Biological Macromolecules Part B* 104:1569–77.

Pepper, R.A., S.J. Couperthwaite, and G.J. Millar. 2017. Value adding red mud waste: High performance iron oxide adsorbent for removal of fluoride. *Journal of Environmental Chemical Engineering* 5(3):2200–6.

Peters, O.M., and C.J. Rauter. 1974. Pathways in thioacetamide hydrolysis in aqueous acid: detection by kinetic analysis. *Journal of the Chemical Society. Perkin Transactions Part 2* 15:1832–5.

Pons, M.N., and J. Dodds. 2015. Particle shape characterization by image analysis. In: S. Tarleton (ed.) *Progress in Filtration and Separation*, pp. 609–636. London: Academic Press/Elsevier.

Pretorius, P.J., and P.W. Linder. 2001. The adsorption characteristics of δ-manganese dioxide: A collection of diffuse double layer constants for the adsorption of H^+, Cu^{2+}, Ni^{2+}, Zn^{2+}, Cd^{2+} and Pb^{2+}. *Applied Geochemistry* 16(7):1067–82.

Qiao, S.-Z., J. Liu, and G.Q. Max Lu. 2017. Synthetic chemistry of nanomaterials. In: R. Xu, and Y. Xu (eds.) *Modern Inorganic Synthetic Chemistry*, 2nd Edn, pp. 613–640. Amsterdam: Elsevier.

Qiu, W., M. Xu, X. Yang, F. Chen, Y. Nan, and H. Chen. 2011. Novel hierarchical CdS crystals by an amino acid mediated hydrothermal process. *Journal of Alloys and Compounds* 509:8413–20.

Rakovich, A.Y., V. Stockhausen, A.S. Susha, S. Sapra, and A.L. Rogach. 2008. Decorated wires as a reaction product of the microwave-assisted synthesis of CdSe in the presence of glycine. *Colloids and Surfaces A: Physicochemical and Engineering Aspects* 317:737–41.

Rantala, T.T. 1999. Ab initio studies of compound semiconductor surfaces. In: Th.F. George, and D.A. Jelski (eds.), *Computational Studies Of New Materials*, pp. 6–26. Singapore: World Scientific Publishing Co.

Rasmussen, K., H. Rauscher, A. Mech, et al. 2018. Physico-chemical properties of manufactured nanomaterials – Characterisation and relevant methods. An outlook based on the OECD Testing Programme. *Regulatory Toxicology and Pharmacology* 92:8–28.

Ronco, C., and J.F. Winchester (eds.). 2001. *Dialysis, Dialyzers and Sorbents: Where Are We Going?* (Contributions to Nephrology, Vol. 133). Basel: Karger.

Roth, H.-C., S.P. Schwaminger, M. Schindler, F.E. Wagner, and S. Berensmeier. 2015. Influencing factors in the co-precipitation process of superparamagnetic iron oxide nano particles: a model based study. *Journal of Magnetism and Magnetic Materials* 377:81–9.

Saha, J.K., and J. Podder. 2011. Crystallization of zinc sulphate single crystals and its structural, thermal and optical characterization. *Journal of Bangladesh Academy of Sciences* 35(2):203–10.

Sampanthar, J.T., J. Dou, G.G. Joo, W. Effendi, and Q.H.E. Low. 2007. Template-free low temperature hydrothermal synthesis and characterization of rod-shaped manganese oxyhydroxides and manganese oxides. *Nanotechnology* 18(2):025601.

Singh, M., D.N. Thanh, P. Ulbrich, N. Strnadová, and F. Štěpánek. 2010. Synthesis, characterization and study of arsenate adsorption from aqueous solution by α- and δ-phase manganese dioxide nanoadsorbents. *Journal of Solid State Chemistry* 183(12):2979–86.

Sofronov, D.S., K.N. Belikov, N.N. Kamneva, E.Yu. Bryleva, and A.V. Bulgakova. 2014. Synthesis of submicron ZnS particles and their sorption characteristics. *Sorption and Chromatographic Processes* 14(1):159–65. (in Russian).

Sofronov, D.S., K.N. Belikov, E.M. Sofronova, P.V. Mateichenko, and N.V. Babayevskaya. 2013. Production of disperse particles of CdS from thiourea solutions in the presence of amino acids. *Functional Materials* 20(1):118–22.

Sofronov, D.S., N.N. Kamneva, A.V. Bulgakova, et al. 2013. Effect of anions and medium pH on the formation of ZnS micro- and nanoparticles from thiourea solutions. *Journal of Biological Physics and Chemistry* 13:85–9.

Sofronov, D.S., N.N. Kamneva, K.A. Katrunov, et al. 2014. Effect of precipitation conditions on the particle size and optical properties of ZnS. *Inorganic Materials* 50(7):817–21.

Sofronov, D., A. Krasnopyorova, N. Efimova, et al. 2019. Extraction of radionuclides of cerium, europium, cobalt and strontium with Mn_3O_4, MnO_2, and MnOOH sorbents. *Process Safety and Environmental Protection* 125:157–63.

Sofronov, D.S., A.M. Odnovolova, L.V. Gudzenko, et al. 2017. Study of Mn^{2+} and MnO^{4-} products interaction in alkaline solution. *Functional Materials* 24(2):322–7.

Sofronov, D.S., E.M. Sofronova, V.N. Baumer, K.A. Kudin, O.M. Vovk, and P.V. Mateychenko. 2011. Formation of nano- and microparticles CdS from thiourea solutions. *Functional materials* 18(4):523–8.

Sofronov, D.S., E.M. Sofronova, V.V. Starikov, et al. 2013. Microwave synthesis of ZnSe. *Journal of Materials Engineering and Performance* 22(6):1637–41.

Song, H., Y.-M. Leem, B.-G. Kim, and Y.-T. Yu. 2008. Synthesis and fluorescence properties of pure and metal-doped spherical ZnS particles from EDTA–metal complexes. *Journal of Physics and Chemistry of Solids* 69:153–60.

Stone, A.T., and J.J. Morgan. 1984. Reduction and dissolution of manganese(III) and manganese(IV) oxides by organics: 2. Survey of the reactivity of organics. *Environmental Science & Technology* 18(8):617–24.

Su, Y.M., Ch.-Y. Huang, Y.-P. Chyou, and K. Svoboda. 2017. Sulfidation/regeneration multi-cyclic testing of Fe2O3/Al2O3 sorbents for the high-temperature removal of hydrogen sulfide. *Journal of the Taiwan Institute of Chemical Engineers* 74:89–95.

Sun, J.Q., X.-P. Shen, K.-M. Chen, Q. Liu, W. Liu. 2008. Low-temperature synthesis of hexagonal ZnS nanoparticles by a facile microwave-assisted single-source method. *Solid State Communications* 147:501–4.

Syamchand, S.S., and G. Sony. 2015. Europium enabled luminescent nanoparticles for biomedical applications. *Journal of Luminescence* 165:190–215.

Taffarel, S.R. and J. Rubio. 2010. Removal of Mn^{2+} from aqueous solution by manganese oxide coated zeolite. *Minerals Engineering* 23:1131–8.

Takematsu, N. 1979–1980. Sorption of transition metals on manganese and iron oxides, and silicate minerals. *Journal of the Oceanographical Society of Japan* 35(1):36–42.

Thirunavukkarasu, O.S., T. Viraraghavan, and K.S. Subramanian. 2003. Arsenic removal from drinking water using iron oxide-coated sand. *Water, Air, and Soil Pollution* 142(1):95–111.

Thuy, U.Th.D., N.Q. Liem, Ch.M.A. Parlett, G.M. Lalev, and K. Wilson. 2014. Synthesis of CuS and CuS/ZnS core/shell nanocrystals for photocatalytic degradation of dyes under visible light. *Catalysis Communications* 44:62–7.

Tiquia-Arashiro, S. and D.F. Rodrigues. 2016. *Extremophiles: Applications in Nanotechnology.* Cham: Springer.

Tombácz, E., R. Turcu, V. Socoliuc, and L. Vékás. 2015. Magnetic iron oxide nanoparticles: Recent trends in design and synthesis of magnetoresponsive nanosystems. *Biochemical and Biophysical Research Communications* 468(3):442–53.

Tripathi, B., Y.K. Vijay, S. Wate, F. Singh, and D.K. Avasthi. 2007. Synthesis and luminescence properties of manganese-doped ZnS nanocrystals. *Solid-State Electronics* 51:81–4.

Turner, A., S.M. Le Roux, and G.E. Millward. 2008. Adsorption of cadmium to iron and manganese oxides during estuarine mixing. *Marine Chemistry* 108(1–2):77–84.

Ul-Haq, I., and F. Haider. 2010. Synthesis and characterization of uniform fine particles of iron(III) hydroxide/oxide. *Journal of the Chinese Chemical Society* 57(2):174–9.

Vadiraj, K.T. and Sh.L. Belagali. 2017. Photoluminescence behavior of manganese doped zinc sulphide, synthesized by hydrothermal process. *Materials Today: Proceedings Part 3* 4(11):11696–9.

Vardhan, K.H., P.S. Kumar, and R.C. Panda. 2019. A review on heavy metal pollution, toxicity and remedial measures: Current trends and future perspectives. *Journal of Molecular Liquids* 290:111197.

Vellingiri, K., K-H. Kim, A. Pournara, and A. Deep. 2018. Towards high-efficiency sorptive capture of radionuclides in solution and gas. *Progress in Materials Science* 94:1–67.

Wang, C., Q. Li, and B. Hu. 2017. Preparation and characterization of ZnS nanoparticles prepared by hydrothermal method. *International Journal of Modern Physics B* 31(16–19):6 p. Article 1744055

Wang, M., L. Sun, X. Fu, Ch. Liao, and Ch. Yan. 2000. Synthesis and optical properties of ZnS:Cu(II) nanoparticles. *Solid State Communications* 115:493–496.

Wang, M., Q. Zhang, W. Hao, and Zh.-X. Sun. 2011. Surface stoichiometry of zinc sulfide and its effect on the adsorption behaviors of xanthate. *Chemistry Central Journal* 5:73.

Wang, X., Y. Zhong, T. Zhai, et al. 2011. Multishelled Co_3O_4-Fe_3O_4 hollow spheres with even magnetic phase distribution: Synthesis, magnetic properties and their application in water treatment. *Journal of Materials Chemistry* 21:17680–7.

Wang, X.F., J.-J. Xu, and H.-Y. Chen. 2008. A new electrochemiluminescence emission of Mn^{2+}-doped ZnS nanocrystals in aqueous solution. *Journal of Physical Chemistry C* 112 (45):17581–5.

Wang, Y., J. Li, J. Li, et al. 2018. Solution prepared O-doped ZnS nanocrystals: Structure characterization, energy level engineering and interfacial application in polymer solar cells. *Solar Energy* 160:353–9.

Watson, J.H.P., B.A. Cressey, A.P. Roberts, et al. 2000. Structural and magnetic studies on heavy-metal-adsorbing iron sulphide nanoparticles produced by sulphate-reducing bacteria. *Journal of Magnetism and Magnetic Materials* 214:13–30.

Wegmann, M., and M. Scharr. 2018. Synthesis of magnetic iron oxide nanoparticles. In: H.-P. Deigner, and M. Kohl (eds.) *Precision Medicine: Tools and Quantitative Approaches,* pp. 145–181. London: Elsevier Academic Press.

Wershin, P., M.H.R. Person, G. Redden, et al. 1994. Interaction between aqueous uranium (VI) and sulfide minerals: Spectroscopic evidence for sorption and reduction. *Geochimica et Cosmochimica Acta* 58(13):2829–43.

WHO (World Health Organization). 2003. Iron in Drinking-water. Geneva. http://www.who.int/water_sanitation_health/dwq/chemicals/iron.pdf.

Wilkin, R.T., and D.G. Beak. 2017. Uptake of nickel by synthetic mackinawite. *Chemical Geology* 462:15–29.

Wu, R., J. Qu, and Y. Chen. 2005. Magnetic powder MnO–Fe$_2$O$_3$ composite—a novel material for the removal of azo-dye from water. *Water Research* 39(4):630–8.

Wu, Y., B. Tang, H. Huo, et al. 2013. The study of zinc sulphide scintillator for fast neutron radiography. *Physics Procedia* 43:205–15.

Xiao, Q., and Ch. Xiao. 2008. Synthesis and photoluminescence of water-soluble Mn^{2+}-doped ZnS quantum dots. *Applied Surface Science* 254:6432–5.

Xie, X., X. Jiang, T. Zhang, and Zh. Huang. 2020. Study on impact of electricity production on regional water resource in China by water footprint. *Renewable Energy* 152:165–78.

Xiong, Z., F. He, D. Zhao, and M.O. Barnett. 2009. Immobilization of mercury in sediment using stabilized iron sulfide nanoparticles. *Water Research* 43:5171–9.

Xu, J., Z. Qu, N. Yan, et al. 2016. Size-dependent nanocrystal sorbent for copper removal from water. *Chemical Engineering Journal* 284:565–70.

Yan, X., E. Michael, S. Komarneni, J.R. Brownson, and Z.-F. Yan. 2013. Microwave- and conventional-hydrothermal synthesis of CuS, SnS and ZnS: Optical properties. *Ceramics International* 39(5):4757–63.

Yao, J., G. Zhao, D. Wang, and G. Han. 2005. Solvothermal synthesis and characterization of CdS nanowires/PVA composite films. *Materials Letters* 59:3652–5.

Yao, Q.Z., G. Jin, and G.-T. Zhou. 2008. Formation of hierarchical nanospheres of ZnS induced by microwave irradiation: A highlighted assembly mechanism. *Materials Chemistry and Physics* 109:164–8.

Zhai, T., Zh. Gu, Y. Ma, W. Yang, L. Zhao, and J. Yao. 2006. Synthesis of ordered ZnS nanotubes by MOCVD-template method. *Materials Chemistry and Physics* 100:281–4.

Zhai, X., X. Zhang, Sh. Chen, W. Yang, and Zh. Gong. 2012. Oleylamine as solvent and stabilizer to synthesize shape-controlled ZnS nanocrystals with good optical properties. *Colloids and Surfaces A: Physicochemical and Engineering Aspects* 409:126–9.

Zhang, Y., F. Lu, Zh. Wang, et al. 2007. ZnS nanoparticle-assisted synthesis and optical properties of ZnS nanotowers. *Crystal Growth & Design* 7(8):1459–62.

Zhang, Y.C., G.Y. Wang, X.Y. Hu, and W.W. Chen. 2006. Solvothermal synthesis of uniform hexagonal-phase ZnS nanorods using a single-source molecular precursor. *Materials Research Bulletin* 41:1817–24.

Zhang, Y.C., G.Y. Wang, X.Y. Hu, Q.F. Shi, T. Qiao, and Y. Yang. 2005. Phase-controlled synthesis of ZnS nanocrystallites by mild solvothermal decomposition of an air-stable single-source molecular precursor. *Journal of Crystal Growth* 284:554–60.

Zhao, Y., J.-M. Hong, and J.-J. Zhu. 2004. Microwave-assisted self-assembled ZnS nanoballs. *Journal of Crystal Growth* 270:438–45.

Zhu, D., J. Zhang, J. Song, et al. 2013. Efficient one-pot synthesis of hierarchical flower-like α-Fe$_2$O$_3$ hollow spheres with excellent adsorption performance for water treatment. *Applied Surface Science* 284(1):855–61.

Zhu, J., M. Zhou, J. Xu, and X. Liao. 2001. Preparation of CdS and ZnS nanoparticles using microwave irradiation. *Materials Letters* 47:25–9.

Zhu, Zh., H. Ma, R. Zhang, Y. Ge, and J. Zhao. 2007. Removal of cadmium using MnO_2 loaded D301 resin. *Journal of Environmental Sciences* 19(6):652–6.

Zhuo, R.F., H.T. Feng, D. Yan, et al. 2008. Rapid growth and photoluminescence properties of doped ZnS one-dimensional nanostructures. *Journal of Crystal Growth* 310:3240–6.

Zou, W., R. Han, Z. Chen, J. Shi, and H. Liu. 2006. Characterization and properties of manganese oxide coated zeolite as adsorbent for removal of copper(II) and lead(II) ions from solution. *Journal of Chemical & Engineering Data* 51(2):534–41.

This chapter is made Open Access through funding by Precision Machine Parts Poland Sp. z o.o.

Index

agglomerate 6, 9, 15, 16, 21, 23, 28, 29, 31, 32, 33, 37, 40, 42, 45, 50, 66, 69–72, 74, 82, 97
anion 26, 28, 48, 55, 56, 71, 72, 97

birnessite 81

cadmium 13, 29–30, 55, 60
   sulfide 2, 3, 14–19, 24, 26, 29, 37, 40, 44–42, 46, 55, 56, 58, 59, 66
   nitrate 20
cation 23, 55, 89
cerium 45, 51, 52, 56, 60, 63, 66, 89–91, 93, 94
   oxide 50, 51
   salt 50
cobalt 8, 15, 56, 57, 59, 62, 63, 64, 66, 68, 89, 90, 91, 93–98
copper 1, 8, 13, 45, 46, 50, 51, 55–57, 62, 63, 66, 68, 89, 90
   sulfide 2, 3, 14–17, 23, 26, 28, 37, 40, 42, 46, 47, 55–61, 63, 64, 66
core-shell 5, 46
covellite 17, 37, 46

europium 45, 50, 52, 53, 55, 56, 66, 89–91, 93, 95, 96, 98
   oxide 51–53

hawleyite 37
hematite 3, 54, 56, 69, 71, 72, 97

ion 6, 9, 12, 20, 21, 26, 45, 46, 48, 51, 55, 56, 58, 59, 62–64, 67, 68, 73, 76, 78–80, 83, 87–89, 91–93, 95, 97, 99
iron 14, 54, 56, 57, 62, 66, 76–80, 89, 90, 97
   chloride 71, 72, 78
   fluoride 73
   hydroxide 68, 71
   nitrate 69, 72–74, 97
   oxalate 74, 75
   oxide 2, 3, 54, 67, 68, 70–73, 75, 76, 78, 79, 88–92, 97–99

stearate 76
sulfate 79
sulfide 14, 54–57, 59, 62, 63

magnetite 3, 76, 77, 78, 79, 80, 97
manganese 1, 8, 45, 48, 49, 51, 56, 57, 63, 64, 81–86, 95
   chloride 86
   hydroxide 50
   nitrate 87, 94
   oxide 50, 59, 60, 62, 63, 68, 80, 81, 86, 87, 91, 97, 98
   oxyhydroxide 2, 3, 81, 85–87, 91, 93, 94, 96–98
   pentahydrate 80
   sulfate 49
   sulfide 48, 50
microwave 2, 5, 7, 38, 46–48, 51–53, 58, 65, 80
monolayer 11, 59, 82

nanoparticle 1, 2, 5, 6, 8–11, 23, 29, 37, 54, 63, 66–68, 70, 71, 75, 76, 78, 80, 87, 97
nanostructure 5, 63, 66–68, 80

potassium
   bromide 7
   hydroxide 20
   permanganate 86
pyrite 55

sodium
   citrate 10
   hydroxide 13, 20, 23, 39, 40
   polyphosphate 11
   sulfide 10, 11, 15, 16, 49, 50, 54, 64, 65
sorption capacity 1, 2, 55, 56, 58–68, 90, 93–99
specific surface 11, 58, 61, 67, 93, 94
sphalerite 6, 10, 17, 18, 37, 46
strontium 56, 57, 89, 91, 93, 95, 96, 98

wurtzite 9, 18

yield 15, 16, 17, 18, 20, 24, 25, 39

zinc 5, 6, 7, 8, 13, 55, 56, 57, 68, 89
    acetate 9
    chloride 10, 26
    diethyldithiocarbamate 8, 9
    dithiocarbamate 8
    ethylenediaminetetraacetate 10
    hydroxide 18
    nitrate 9, 13, 48
    oxide 1, 17, 18
    stearate 8
    sulfate 9
    sulfide 1, 2, 3, 5, 6, 7, 8, 9, 10, 14, 15, 17, 18, 19, 20, 21, 23, 24, 26, 37, 39, 40, 42, 44, 45, 46, 48, 50, 51, 58, 59, 60, 63, 64, 65, 66

Printed in the United States
By Bookmasters